Elements of BIOTECHNOLOGY

Promises and Realities

Elements of BIOTECHNOLOGY

Promises and Realities

Sidharth Baliyan

ANMOL PUBLICATIONS PVT. LTD.
NEW DELHI-110 002 (INDIA)

ANMOL PUBLICATIONS PVT. LTD.
Regd. Office: 4360/4, Ansari Road, Daryaganj,
New Delhi-110002 (India)
Tel.: 23278000, 23261597, 23286875, 23255577
Fax: 91-11-23280289
Email: anmolpub@gmail.com
Visit us at: www.anmolpublications.com

Branch Office: No. 1015, Ist Main Road, BSK IIIrd Stage
IIIrd Phase, IIIrd Block, Bengaluru-560 085 (India)
Tel.: 080-41723429 • Fax: 080-26723604
Email: anmolpublicationsbangalore@gmail.com

Elements of Biotechnology: Promises and Realities

First Edition, 2011

PRINTED IN INDIA

Printed at Mehra Offset Press, Delhi.

Contents

Preface

Biotechnology is an international peer-reviewed scientific journal which publishes outstanding research papers from all areas related to biotechnology. Scope of biotechnology covers: molecular biology, food biotechnology, genetic engineering, microbial biotechnology, plant biotechnology, animal biotechnology, marine biotechnology, industrial biotechnology, environmental biotechnology, biological processes, industrial applications and bioinformatics. Biotechnology in one form or another has flourished since prehistoric times.

When the first human beings realized that they could plant their own crops and breed their own animals, they learned to use biotechnology. The discovery that fruit juices fermented into wine, or that milk could be converted into cheese or yogurt, or that beer could be made by fermenting solutions of malt and hops began the study of biotechnology. In past few years biotechnology and its whole field coped up with the development in our country. Not only our country developed its biotechnology but by importing means of development of the said field of biotechnologies also made the difference in the same.

Author

1

Metabolism

INTRODUCTION

The term ADME is typically used in reference to nonclinical studies. In reality, ADME is generic, being applicable without restriction to pharmacokinetic/metabolism investigations in humans as well as animals. Fundamentally, ADME information is critical in all phases of a fully integrated drug development programme. The initial charge of early development programmes is to file the Investigational New Drug (IND) application such that approval may be secured to investigate a new chemical entity (NCE) in humans.

Animal toxicology studies comprise the foundation of the IND. The toxicology programme is designed to identify toxicities that may potentially be encountered in humans, helping ensure that the initial human studies are conducted safely and ethically. In this context, ADME studies provide supportive information to augment the interpretation of toxicology findings. Of primary importance among them are toxicokinetic studies, in which systemic drug and/or metabolite exposure in toxicology animals is evaluated. Drug exposure, expressed in terms of AUC, Cmax (maximum drug concentration in plasma), or an alternative parameter, is then related to dose level and toxicological outcomes. Based on toxicokinetic data at the no-observed toxic effect dose, an acceptable exposure limit in humans can be defined. To assist in putting toxicokinetic data into a broader perspective, the basic pharmacokinetic behaviour of the NCE is assessed in the toxicology species. Some typical studies that may be conducted in this stage of development are listed in table.

Table. Adme Studies Which Support an Ind.

Toxicokinetics
Pharmacokinetics/absolute bioavailability in toxicology species (male/female)
Protein binding
Erythrocyte/plasma distribution
Whole body autoradiography/tissue distribution
Mass balance in toxicology species
Metabolite profile in toxicology species
Pharmacodynamics
Allometric scaling

In later development, the focus of ADME work is shifted to human studies. These are intended to more fully define the disposition of the drug in humans, particularly in the target therapeutic population. Ultimately these data are integrated into a New Drug Application (NDA) or Market Authorization Application (MAA) to secure final approval to market the drug. Possible studies, which may be performed during clinical development, are listed in Table.

Table. Adme Studies Which Support an NDA/MAA.

NONCLINICAL
Toxicokinetics in chronic and reproductive/teratology studies
Multiple dose pharmacokinetics in toxicology species
Biliary excretion/enterohepatic recirculation
Metabolite identification in toxicology species
Multiple dose whole-body autoradiography/tissue distribution
Placental transfer
Milk secretion
Effects on metabolic systems (induction)
CLINICAL
Single/multiple dose pharmacokinetics in safety/tolerance studies
Dose proportionality
Food effect
Repeated measures (intrasubject variability)
Mass balance/metabolite profile and identification with radiotracer
Pharmacokinetics in subpopulations (gender, age, genetics, liver and renal failure)
Drug interactions
Milk penetration
Pharmacodynamics
Population pharmacokinetics/pharmacodynamics in patients
Pivotal bioequivalence

ADME data gathered during the full course of development are incorporated into drug labeling, which is intended to optimize therapeutic utilization of the drug in the target population

ADME IN THE DRUG DISCOVERY PROCESS

While scientists in the ADME disciplines have historically provided support to drug discovery teams, the unique expertise, which they could bring to the discovery process, has heretofore been vastly underutilized. In no small way this may be ascribed to the less than expedient experimental procedures required to conduct most ADME studies. The time- and labour-intensiveness of much of this work is not an issue during drug development, where resources can be focused on one compound. This *modus operandi* does not conform, however, to current drug discovery efforts, which have exponentially enhanced their operating efficiency with mass-screening techniques. However, the advent of a host of exciting new technologies is providing hope that scientists in various of the ADME disciplines may now be able to participate more actively and productively in the drug discovery process.

Absorption

Insofar as the overwhelming majority of new drugs are intended to be administered orally, with few exceptions the ability of an NCE to provide activity by the oral route (*i.e.* have good bioavailability) is imperative. Whereas oral bioavailability can be determined relatively easily in animal pharmacokinetic studies, it is not reasonable to expect that this approach can function optimally in a large scale-screening environment. Fortunately a number of *in vitro* and cell culture techniques have evolved in recent years that have facilitated the assessment of intestinal permeability.

It has been demonstrated that membrane permeability can be predicted for some compounds with reasonable accuracy based solely on physicochemical parameters. Therefore, close scrutiny of the chemical structure may provide valuable basic information about an NCE prior to the commencement of laboratory experiments. It is well established, for instance, that efficient oral absorption will occur only after drug has dissolved and presented itself to the intestinal mucosal surface from whence it can traverse the epithelium. Dissolution is determined by the highly interdependent influences of aqueous solubility, ionizability (pKa), and lipophilicity (octanol/water log P or log D7.4). Furthermore, log P is a crucial factor governing passive membrane partitioning, influencing permeability opposite to its effect on solubility (*i.e.* increasing log P enhances permeability while reducing solubility). In light of this counterdependence, it has been suggested that oral absorption may be optimal within a log P range of 0.5 to 2.0.

While the promise of predicting membrane permeability from chemical structure alone is enticing, these methods do not yet enjoy the level of sophistication required to supplant experimental methods. In that regard, a number of *in vitro* tools have been adapted, with notable efficiency, to high-throughput assessments of membrane permeability and potential oral bioavailability. Most notable among them are CaCO-2 cells. Derived from a human colon carcinoma cell line, these cells are grown in a confluent monolayer on porous membrane filters which are mounted in diffusion chambers. Permeability measurements are based on the rate of appearance of test compound in the receiver compartment. The apical (donor) surface of the monolayer contains microvilli and thus retains many characteristics of the intestinal brush border. The cells also express functional transport proteins and metabolic enzymes, the degree of expression being dependent upon the post-seeding age of the cells.

Everted intestinal rings and brush-border membrane vesicles (BBMV) are also commonly used systems for assessing membrane permeability. The former technique is a refinement of one of the earliest *in vitro*absorption systems in which an everted intestinal segment was suspended in a buffer system to measure mucosal-to-serosal transfer. The current methodology involves isolating a rat intestinal segment, everting, and slicing into rings, which are suspended in buffer. Removing the brush-border surface from rat or rabbit intestine and molding it into vesicles by homogenization and differential centrifugation prepare BBMV. Both everted intestinal rings and BBMV are most useful for determining drug uptake rates rather than transepithelial flux.

In the *in situ* intestinal perfusion system, an intestinal segment is exposed in an anesthetized rat and drug solution is perfused through the lumen in a single-pass or recirculating fashion. Drug permeability is derived from the rate of disappearance of drug from the perfusate. Though more labour-intensive, in situ intestinal perfusions remain popular owing to the perceived clinical relevance of permeability data derived therefrom. In a recent study, it was demonstrated with a series of small organic molecules as well as a series of peptidomimetics that CaCO-2 cells, everted intestinal rings, and in situ perfusions have strong potential for predicting fraction absorbed in humans.

One of the most appealing attributes, which these experimental systems possess, is their capability to perform relatively high throughput screening. This is particularly relevant for CaCO-2 cells, everted intestinal rings, and BBMV. An imposing rate-limiting step, once the systems are established and optimized, is the development of assay methods (usually HPLC) to quantify the analytes of interest. On the positive side, as these experiments are conducted in aqueous buffers, the preanalytical sample purification requirement is minimal. These analytical burdens can be even further reduced by immobilized artificial membranes (IAM), a recently developed system which has been greeted with considerable enthusiasm. Briefly, the analyte is injected onto a specialized IAM HPLC column packed with a phosphatidylcholine (PC) stationary phase selected to closely mimic biological membranes.

Theoretically, the chromatographic capacity factor for the analyte should correlate with the water-to-PC membrane partition coefficient, a useful parameter for calculating membrane permeability. The ability of IAM technology to predict membrane permeability is currently being evaluated. Elegant in its simplicity, it is well suited to a rapid screening environment, although it is recognized that its applicability will be limited to those compounds, which are absorbed by passive processes.

Metabolism

The systems described above have the greatest utility when absorption is rate limiting to systemic availability. For many compounds, even if absorption is optimized bioavailability may be limited by extensive metabolism. Indeed, metabolism can complicate the *in vivo* activity profile irrespective of route of administration.

As with absorption, valuable metabolic input can be imparted to the drug discovery team based on information gleaned from chemical structure alone. It is felt that the teleological underpinnings of xenobiotic transformation reside in modifying environmental toxins so as to facilitate their removal from the body. Not surprisingly, the majority of drug biotransformation processes appear to follow this paradigm, typically resulting in increasingly water-soluble metabolic products which can be more readily excreted in urine. Thus one may surmise that within a group of compounds, highly lipophilic drugs would be metabolized most actively.

More explicitly, the metabolic potential of a compound containing a chemical moiety known to be avidly biotransformed may be predicted reasonably accurately on the basis of abundant historical data. For example, being cognizant that unhindered phenolic hydroxyl groups are exquisitely good substrates for conjugating enzymes, NCEs, which contain this moiety, can be eliminated from further consideration if a predilection to rapid metabolism is felt to be detrimental. As we gain information about the specific conformational constraints imposed by the catalytic sites of major metabolic enzymes, our ability to predict *in vivo* metabolic events based solely on chemical structure is enhanced commensurately.

Given the current state of knowledge, however, it is likely that drug discovery teams will continue to prefer metabolism-related input predicated on experimental data rather than theoretical predictions. To this end, various *in vitro* methods are available which are being increasingly incorporated into drug discovery strategies. Among the most popular and widely utilized systems in use today are hepatic microsomes. These preparations retain activity of those enzymes, which reside in the smooth endoplasmic reticulum, such ascytochromes (CYP), flavin monooxygenases, and glucuronosy ltransferases. Isolated hepatocytes appear to retain a broader spectrum of enzymatic activities, including not only reticular systems, but cytosolic and mitochondrial enzymes as well. Because of a rapid loss of hepatocyte-specific functions, it has been possible to generate useful data only with short term hepatocyte incubations or cultures. However, significant strides have been made towards maximizing the viability of hepatocytes in culture.

Liver slices, which like hepatocytes retain a wide array of enzyme activities, are also increasing in popularity. Furthermore, both hepatocytes and liver slices are capable of assessing of enzyme induction *in vitro*. The choice of which system to employ in a drug discovery screening programme will depend on many factors, not the least of which is availability of tissue for large-scale implementation. In addition, historical information about a particular chemical series is invaluable.

Consider, for example, an SAR-driven chemical series with a prototype analogue that has been demonstrated to induce CYP activity. For metabolic screening in this programme, hepatocytes or liver slices may be preferable to microsomes. Biotransformation of NCE candidates as well as induction potential can be evaluated in the same system, making the added resources required to establish these systems well worth the investment. Information gathered from *in vitro* metabolism studies is especially useful in choosing drug candidates for future development. The potential utility of *in vitro*metabolism data in predicting *in vivo* intrinsic clearance has been touted.

Isolated heterologous human CYP enzymes have been available for several years, being expressed from cDNA in yeast (*Saccharomyces* cerevisiae), bacterial (*Escherichia coli*), and mammalian (B-lymphoblastoid) cell lines. These systems can be used to ascertain whether a compound is a substrate for a particular CYP isozyme and, if so, what metabolite is generated by that enzyme. Moreover, these enzymes, in sufficient quantity, may possibly be used as bioreactors to generate usable amounts of a metabolic product that may be difficult to chemically synthesize in the laboratory. Isozyme-specific antibodies and isozyme-specific inhibitory substrates, by selectively abolishing the activity of a particular CYP isoform, may be used to determine the relative importance of that enzyme in the turnover of an NCE. Conversely, the ability of an NCE to interact with a particular CYP isozyme may be determined by metabolic cross-inhibition studies against prototypical CYP substrates.

In vitro metabolism systems are not limited to those derived from the liver. Most pharmaceutical companies have been building liver and tissue banks to permit a cross-comparison of metabolic turnover rates in various tissues from various species. It is not uncommon now for liver banks to house tissues from a variety of species, including those from animals treated with enzyme inducers or inhibitors. Therefore, cross-species *in vitro* metabolism comparisons are becoming more feasible and commonplace. In addition to providing information on potential rates and routes of metabolism, interspecies comparisons may help in choosing species to be used in toxicology studies.

Available *in vitro* metabolism technologies are considerably more efficient than traditional *in vivo* methods. With these systems, it is possible to assess the relative rates and routes of biotransformation of a handful of compounds in the time required to comparably characterize one compound using *in vivo* approaches. In the past, a relative paucity of tissues was the most notable impediment to conducting large-scale screening efforts with *in vitro* tools. However, the continued growth of in-house tissue banks, in concert with the optimization of isolated enzyme expression systems, has made high-throughput metabolism assessments in the discovery arena a reality.

PHARMACOKINETICS

Unfortunately, there is no substitute for actual *in vivo* data in assessing pharmacokinetic profiles of drug candidates. While insight into various aspects of the pharmacokinetic profile (absorption, metabolism, protein binding) can be gleaned from *in vitro* techniques, there are as yet no methods available for accurately predicting what will happen to a drug when it is put into a whole animal.

For a useful assessment of pharmacokinetics and bioavailabilit~~y, it is~~ necessary to administer the drug to selected animal species both intravenously and by the intended route of administration (usually oral). Blood samples are collected over a predetermined time course after dosing and the drug is quantified in serum or plasma by a suitable bioanalytical method (*e.g.* HPLC). Alternatively or concurrently it may be possible to collect plasma samples from animals used in whole animal pharmacologic models and, based on the concentration/ effect (pharmacokinetic/ pharmacodynamic) relationship established, make a link between *in vitro* pharmacologic activity and the behaviour of a compound *in vivo*.

In any case, the most significant impediment to providing pharmacokinetic input to a high-throughput discovery team is the time and labour-intensiveness of the bioanalytical methods available. To assess biochemical or pharmacologic activity *in vitro* or *in vivo*, a standardized screening method is established with a common assay endpoint that can be applied to test all compounds that are available. In contrast, a separate bioanalytical method for pharmacokinetic assessment must be developed for each compound.

For chromatographic assays, which comprise the vast majority of the methods employed in pharmacokinetic studies, remarkable improvements in assay detection limits and sample cleanup have been realised. However, little has been done to shorten the time required for assay development and implementation. Pharmacokinetic characterization is therefore often relegated to end-stage discovery, being utilized to select which of 2 or 3 potential lead candidates has the most "acceptable" pharmacokinetic profile. Clearly, if pharmacokinetic input is to be available for early discovery decisions, more efficient methodologies are necessary. Semi-simultaneous bioavailability estimation is a screening method which has been successfully utilized in drug discovery. With this technique, an intravenous dose of drug is administered and at a suitable time postdose (*i.e.* postdistributional) an oral dose is administered to the same animal.

Pharmacokinetic parameters such as clearance and volume of distribution can be determined from the intravenous concentration-time curve. Oral pharmacokinetic parameters, including bioavilability, are subsequently extricated from the combined IV and PO concentration-time data by deconvolution. By reducing inter- and intraanimal variability, accurate pharmacokinetic data can be gathered with fewer animals. In addition, the total number of plasma samples from both routes of administration is reduced by the overlapping dosing regimes. On the other hand, a prior knowledge of the pharmacokinetic characteristics of the NCE is needed to optimize the administration paradigm. In that regard, it may be possible to extrapolate from studies with structurally analogous predecessors from the same chemical series. More critical, however, is that compound-specific analytical methods are still required for each NCE.

Bioassays represent one alternative to compound-specific assays. Plasma samples are added to an *in vitro*system and the biological activity is quantified in much the same way as a pure drug solution is tested for activity. The extent of displacement of a prototype receptor ligand or inhibition of a target enzyme activity is measured as the assay endpoint for each sample. Based on standard curves, which relate the assay endpoint to known quantities of added drug, the concentration of the compound in the plasma can be quantified. Microbiological assays, which measure the inhibition of bacterial growth on an agar plate in relation to the amount of antibacterial drug applied, are among the most widely utilized examples of this technique.

The greatest advantage of bioassays over HPLC methods is that they can be employed for analyzing any compound which possesses the appropriate biological activity, making them amenable to high throughput screening. Unfortunately, owing to their non-specific nature, these assays will quantify all biological activity in a sample, which could include active metabolites. These results may be misleading if one is attempting to understand the pharmacokinetic behaviour of only parent compound. In response, it could be argued that it is prudent to choose the compound, which provides the best activity profile, irrespective of whether the in vivo activity observed, is due to parent compound or active metabolite. Preprocessing of plasma samples (extraction) could possibly remove polar metabolites, allowing the bioassay to concentrate on parent drug.

In perhaps the most significant technologic leap in this field, scientists are capitalizing on the exquisite selectivity of mass spectrometry (MS) to circumvent the tedium of HPLC assay development. The resulting procedure is the pharmacokinetic equivalent of a mass-screen, the simplicity and elegance of which is the result of several evolutionary steps. In its earliest incarnation, animals were dosed with a single drug and samples were collected, processed, and injected onto an HPLC- MS. The chromatographic system was optimized to isolate the drug peak and the mass spectrometer was used to detect the appropriate mass ion and quantify it. Eventually, HPLC separation was foregone in favour of direct injection of a minimally preprocessed plasma sample into the MS. In a further refinement, multiple drugs were boldly administered simultaneously to the same animal and each compound was then individually quantified.

MS is capable of distinguishing subtle differences in the characteristic mass ion peaks of each analyte, imparting a degree of specificity that HPLC cannot achieve without fairly elaborate intervention. Under this protocol, the number of drugs for which pharmacokinetic profiles can be obtained in a week increases several-fold, being limited only by the number of compounds, which are co-administered in each animal. The considerable sample burden thus engendered is easily accommodated by the facile, direct-injection MS analysis. Overcoming myriad theoretical reasons why it shouldn't work (*e.g.* potential for metabolic interactions, protein binding displacement, or additive pharmacologic effects), investigators in several large pharmaceutical companies have successfully incorporated this screening approach into routine drug discovery support.

Drug distribution and tissue penetration data are occasionally obtained during discovery screening. Plasma protein binding assessments can be relatively easily performed using well-known and widely-used methods, providing perspective on the relationship between total *in vivo* plasma drug concentrations and those of pharmacologically available unbound drug. For more specific applications, cell cultures techniques are available for predicting the ability of compounds to traverse target biological membranes (*e.g.* blood-brain barrier, T lymphocytes). *In vivo* microdialysis has been used for assessing drug penetration into CNS and liver, but is methodologically difficult to transform into a screening tool. In addition, these methodologies typically suffer from the same bioanalytical limitations of classical pharmacokinetic profiling. It seems reasonable to hope that direct-injection MS analysis may facilitate these types of studies in the future.

ADME IN DRUG DESIGN

With continued refinement and streamlining of its underlying technologies, ADME-relateddiscovery screening can be implemented ever earlier into the discovery process. Ultimately, this will permit scientists in the ADME disciplines to play a significant role inoptimizing new drug design. Fundamentally, that role is to identify the limiting barriers to optimal *in vivo* efficacy.

Of all the attributes, which determine a drug's ultimate *in vivo* efficacy, physicochemical behaviour is perhaps the most basic. For this reason alone it deserves considerable scrutiny by the drug discovery team. Moreover, as solubility, pKa, and lipophilicity are so integrally linked with chemical structure, physicochemical properties in some respects can be relatively predictably manipulated by chemical modifications.

Historically, chemical synthetic efforts have been guided almost exclusively by SAR intended to maximize interaction with a therapeutic target (*e.g.* receptor, enzyme). The goal of ADME-guided synthesis, on the other hand, is to maximize the ability of an NCE to access the therapeutic target *in vivo*. As these synthetic directives may often be at odds, the Holy Grail for discovery teams in the future will be to achieve an optimal balance between the interests of these groups and fully integrate the valuable information which they can each impart towards guiding new drug design.

Lipophilicity is a critical determinant of a host of ADME processes. It has been suggested, for example, that compounds with log P values between 0 and 3 or 4 (*e.g.* 0.5 to 2) are the most suitable candidates for passive transcellular absorption across intestinal epithelia. As lipophilicity is increased (log P greater than 3 or 4), solubility and hence absorption progressively decline.

In contrast, more hydrophilic compounds with log P values less than 0 are likely to traverse the epithelium more slowly via paracellular channels, although it should be appreciated that molecular size is also an important determinant of transit through the narrow (5-9 Å) paracellular channels. Similarly, it has been observed that compounds with log P values greater than 0 are likely to undergo substantial renal tubular reabsorption. By itself, this effect will tend to prolong $t_{1/2}$, however the greaterlipophilicity and minimal renal excretion will also render the compound more susceptible to metabolism.

Experimental determination of lipophilicity is not a trivial matter, a fact which has limited the availability of this information in early discovery where it could be most useful. However, mathematical predictions based on a weighting of functional groups within a molecule have been performed with impressive accuracy. As confidence in the reliability of these predictions is further developed, it will become more feasible in the synthesis of new drugs to target log P values to achieve a balance between particular ADME processes. Even more so than log P, pKa is a physicochemical property that can be readily predicted, given the vast experience amassed over many years with myriad ionizable organic functional groups. The ionization state of an NCE *in vivo* can be predicted by considering its pKa relative to pH in various body compartments. It is possible to surmise, for example, that an organic acid will present to the nephron in an anionic state, rendering it potentially susceptible to active tubular secretion by the organic anion transporter. If the compound is indeed a substrate for this transporter, renal clearance may be high and elimination from the systemic circulation rapid.

Both log P and pKa are the primary determinants of compound solubility. *In vivo,* this has a decisive bearing on oral absorption, following the premise that drugs must dissolve in the GI tract to be available to cross the intestinal membrane. Given that pH progressively increases throughout the length of the gastrointestinal tract, solubility and thus absorbability of an ionizable compound will vary accordingly. Solubility, log P, and pKa are closely intertwined determinants of renal tubular reabsorption as well. Thus, whereas compounds with log P values greater than 0 may be potential candidates for tubular reabsorption, this process will only be likely to occur if the compound is present in urine (pH 4.5 to 8) in an un-ionized state and has not exceeded its solubility limit.

Despite the importance of physicochemical properties in determining the ADME profile, they are as yet primarily rough guidelines and cannot substitute for experimentation to bear out the validity of their predicted *in vivo* effects. Experimental data generated by the discovery tools outlined above can help put the importance of physicochemical properties into the appropriate context.

Structural modifications may also be guided by more fundamental knowledge of a particular ADME process. The stability of peptidomimetic compounds to peptidases in the GI tract may be improved by various synthetic alterations, including protection of peptide bonds with steric bulk or substitution of naturally occurring L-amino acid residues with their D-isomers. The increased GI stability potentially increases bioavailability, but must be balanced against the contribution of physicochemical properties such as lipophilicity. In addition, it is important to note that although lipophilicity and GI stability may favour intestinal absorption, high molecular weight (>600 Da) of some of these compounds could render them acutely sensitive to biliary excretion, reducing absolute systemic bioavailability in spite of their good intestinal permeability.

Among the many transport proteins, which reside in the GI mucosa, the dipeptide transporter has been shown to be an important mediator of active drug absorption. In addition to di- and tripeptides, compounds from several compound classes, which are structurally similar to di- and tripeptides, have also been shown to be substrates for this carrier, including -lactams and ACE inhibitors. Targeting synthesis to enhance the affinity of NCEs for this transporter may prove to be a fruitful approach to improving systemic bioavailability, particularly for polar compounds with inherently low membrane permeabilities.

In one such example, formation of a proline adduct improved the transcellular passage of L-methyldopa by transforming it into a substrate for the dipeptide transporter. In addition to being dipeptide-like, L-methyldopa-pro is also a substrate for intracellular prolidase which was specifically exploited to regenerate L-methyldopa after the adduct was transported across the apical surface of the enterocyte. As their substrate specificities are further delineated, other known transporters could be similarly targeted for enhancing bioavailability.

Prodrug strategies have successfully improved the oral bioavailability of numerous compounds. In many cases, this involves masking a polar group by esterification to increase lipophilicity and enhance the extent of absorption from the GI tract. After absorption, the ester is enzymatically hydrolyzed to release parent drug. Absorption has also been improved with prodrugs designed to enhance solubility by decreasing lipophilicity, targeting progroup cleavage for the GI tract. With this approach, the nature of the progroup predetermines whether parent drug is released in the GI lumen or at the brush border membrane. The ideal candidate for the latter manipulation is a highly lipophilic compound, which readily traverses the GI mucosa, but is too insoluble to produce a reasonable soluble fraction for acceptable bioavailability. A polar progroup (*e.g.* phosphate, amino acid) is incorporated into the molecule, markedly increasing its solubility in the GI tract. The progroup is subsequently cleaved at the intestinal brush border by the targeted enzymes (*e.g.* phosphatases, aminopeptidases), which expels the lipophilic parent compound in close proximity to the intestinal surface. The low local concentration minimizes the possibility of precipitation and allows the drug to be directly absorbed. The bioavailability of several compounds has been improved with this strategy.

Soft drugs provide an interesting counterpoint to prodrugs. In both cases, the administered compound is designed to undergo a predictable metabolic modification. With prodrugs, this biotransformation event rapidly converts an inactive precursor into an active drug. Conversely, a soft drug is itself the active drug, which, after eliciting a therapeutic effect, is metabolically inactivated in a predictable manner and at a controlled rate. Several approaches to rational soft drug design have been identified, including soft analogs, activated soft compounds, natural soft drugs, the active metabolite approach, and the inactive metabolite approach. In the latter case, an inactive metabolite to a known drug is first identified. This metabolite is then chemically modified to create a new drug which:

Possesses comparable pharmacologic activity to the original drug:

- Will be biotransformed to the inactive metabolite in one step and without going through toxic intermediates, and
- May possibly be subjected to further modification if transport or binding properties require optimization.

The metabolic systems typically harnessed by soft drugs are hydrolytic enzymes (*e.g.* esterases). Monoxygenases are scrupulously avoided in the belief that hydrolyses are inestimably more predictable and controllable processes. In reality, it may be difficult to integrate the soft-drug concept into novel, innovative drug discovery programmes. Its retrometabolic philosophy is based on the premise that the metabolic fate of a pharmacologically active drug is known in explicit detail. Therefore, unless the synthetic strategy is predicated on a well-established chemical series, the information required for soft-drug design will not be available. However, the basic principles of this concept might possibly be extrapolated from successful soft drugs to novel compounds with similar functional moieties.

The guidance of drug design based on more classical metabolism theory is multifaceted. As with any strategy, an ability to predict the metabolic fate of an NCE based simply on its structure would be of untold value. To that end, several computerized " Expert systems" have been developed which are designed to promulgate for any given compound a list of likely metabolic products. Unfortunately, these systems have proven to be perhaps too general in their scope, projecting an array of potential products so extensive as to be of relatively limited use. On a slightly different tack, a vast metabolite database is currently being developed which is based on metabolism studies reported in the literature over the past 90 years. This package, from Molecular Design, interfaces with ISIS databases and is thought to hold considerable promise for providing more realistic projections of metabolic routes across species.

Molecular modeling may come to fruition in the near future. Computer assisted drug design might allow the determination of NCE-receptor compatibility to proceed concurrently with an evaluation of the ability of the NCE to fit into the active site of a particular metabolic enzyme. This capacity does not presently exist on any significant scale. It is particularly frustrating with respect to the CYP family, which is the most important of Phase 1 metabolic enzyme systems. In part because CYP is a membrane-bound enzyme in close association with a reductase, it has not yet been crystallized. Therefore, an X-ray structure, required for building a useful computer model, is not available. The multiplicity of the CYP family makes the task even more formidable. Therefore, guiding synthesis towards an optimal metabolic profile is based as yet on largely empirical insight. Substantial inroads have been made in this regard, particularly with respect to 4 of the most important CYP isozymes, CYP2D6, 2C9, 2C19, and 3A4. Models of the active sites of CYP2D6 and CYP2C19 have been hypothesized based on a comparison of chemical structures of a variety of substrates, taking into consideration the strict stereo- and regioselectivity of their transformations. The high degree of selectivity inherent in their active sites make it feasible to design drugs which may be more or less likely to interact with these isozymes.

In one example, addition of a bulky para-substituent to the basic beta-blocker nucleus ultimately yielded betaxolol, which unlike its pharmacologic brethren is resistant to the action of CYP2D6. CYP3A4, on the other hand, has to date defied attempts to define its active site, being far less discriminating in its choice of substrates. Indeed, some rather large molecules, such as cyclosporin and macrolide antibiotics are known to be turned over by CYP3A4.

As metabolism databases expand in the coming years, a better delineation of how to chemically modify NCEs will be forthcoming, permitting more predictable interactions with CYP isozymes. Some general rules are in place with regard to avoiding chemically reactive structures. Thus, compounds containing structural elements with a high risk of producing toxic products should be strictly avoided. Examples include quinones and quinonimines, aromatic methylene-dioxy groups, and aromatic nitro groups.

Enzyme induction is more difficult to predict. There is little guidance with regard to what chemical structures are known to produce this effect, with the possible exception of broad chemical groups such as barbiturates, hydantoins, di-t-butyl phenols, etc. However, even within such series, subtle changes in chemical structure can markedly and unpredictably alter enzyme induction potential . As detailed above, pharmacokinetic characterization is often relegated to end-stage discovery, at which point compound attrition has whittled the pool of potential lead compounds to a small, manageable number. Under these circumstances, the opportunity to provide meaningful feedback, which might assist in pharmacokinetically optimizing drug design, is limited. As analytical techniques become more expeditious, pharmacokineticists will play an increasingly important role in drug design. Nonetheless, it should be clearly recognized that pharmacokinetics in and of itself is not particularly meaningful.

Pharmacokinetic behaviour is highly interdependent with absorption and metabolism. However, characterizing the pharmacokinetic profile of a compound is still of utmost importance insofar as it helps put absorption, metabolism, and other data into the proper perspective. For example, *in vivo* activity may be poor despite predictions from *in vitro*absorption and metabolism studies that an NCE will be well absorbed and not be subjected to appreciable biotransformation.

The pharmacokinetic profile reveals that the drug has a short *in vivo* half-life owing to a small volume of distribution and high renal clearance, neither of which could have been predicted accurately prior to the whole animal work. In this case, an optimization of the PK profile may be guided by a realization that chemical modifications are needed which will specifically reduce renal clearance or alter plasma protein binding.

The ability to predict *a priori* the extent of plasma protein binding is limited, although fruitful attempts have been made to define the molecular binding requirements for certain sites on human serum albumin. In general, binding is likely to increase with lipophilicity; an anionic group may enhance binding to albumin even further. In contrast, cationic molecules are virtually precluded from binding to albumin, rather favouring -1-acid glycoprotein. Ultimately, despite formidable theoretical arguments to the contrary, the role of protein binding in determining *in vivo* efficacy is still not universally predictable. For example, it has been suggested that NCEs with low protein binding are the most ideal candidates for further development. Quite often this implies low lipophilicity, which could be at odds with strategies for optimizing membrane permeability or receptor or enzyme affinity. It has also been suggested that high protein binding is not necessarily a negative attribute and in fact may help prolong the activity of some compounds with otherwise short half-lives.

2

Ionisation and Buffering

WATER

Sometimes the hydrogen of one water molecule will "jump" to another water molecule:

$$H_2O + H_2O = H_3O^+ + OH^-$$

This proton hopping is called the ionization of water (an ion is a positively or negatively charged atom or molecule). This ionization creates a H_3O^+ and a OH^- molecule. The H_3O^+ is often written as simply H^+. This is because a H_3O^+ is just a H^+ that jumps from one water molecule to another:

$$H_2O = H^+ + OH^-$$

So remember, $H_3O^+ = H^+$

Looking at either of the two chemical equations above, it is important to note that the reverse reaction is also occurring.

How much H^+ and OH^- exist in water? Very, very little! The ratio of either H^+ or OH^- to H_2O in neutral water is 1:1,000,000,000! Since this is such a small amount of either H^+ or OH^-, they rarely meet and neutralize each other.

The equilibrium constant, K_{eq} describes the ionization equilibrium of water:

$$K_{eq} = [H^+][OH^-]$$

Because of this relationship it is important to note that if the $[H^+]$ goes up then the $[OH^-]$ must go down, and vice-versa, for the value for the K_{eq} of water must remain constant. For neutral water, the K_{eq} is 1×10^{-14} M and the concentrations of $[H^+]$ and $[OH^-]$ are each 1×10^{-7} M. Let's look at that last number without the exponent:

$$0.0000001\ M$$

This is obviously a very small number. A more manageable way to discuss small numbers such as this is to take the negative logarithm. For the concentration of $[H^+]$, this is called the pH. In this case:

$$-\log(0.0000001\ M) = 7$$

The pH of a solution is simply the negative logarithm of $[H^+]$. The pH of a solution describes the acidity of a solution. Acidic solutions are those with a pH of less than 7 and basic solutions have a pH greater than 7. A solution, like H_2O, with a pH = 7 is neutral. Similarly, the pH could be used to describe a solution in terms of its OH^- concentration. pOH is the negative logarithm of the OH^- concentration.

One useful thing to remember is:

$$pH + pOH = 14.$$

In the body, the pH of blood is 7.4. This corresponds to a $[H^+]$ of about 40 nM. This value can only vary from 37 nM to 43 nM without serious metabolic consequences.

pKa

In living systems, much of the chemistry involves interactions between acids and bases. Acids are H^+ donors andbases are H^+ acceptors.

$$HA{=}H^+ + A^-$$

Acid and base reactions are made up of conjugate acid-base pairs. Strong acids are those that readily give up a H^+. Strong bases readily accept a H^+. Weak acids do not readily give up a H^+, but will under the right conditions. A weak base is one that does not easily take up a H^+.

The conjugate base for a strong acid is a weak base. In contrast, a strong base has a weak acid as its conjugate. Think of a strong acid as a person with an ugly wig. This person can't wait to get rid of this! So now we have a wig (H^+) and a person without the wig (the weak base). You can understand that the person without the wig doesn't really want to take it back!

However, what about strong base-weak acid conjugate pairs?

$$A^- + H^+ \rightleftharpoons HA$$

Think of it in similar terms: a person in dire need of a wig will grab just about anything. This is our strong base. Now on the right side of the equation we have someone with a wig who really doesn't want to give it up, just as a weak acid does not want to part with its H^+.

We can think of a strong acid or base as having a "flip side", which is its weak base or weak acid conjugate, respectively.

How readily an acid gives up its H^+ is expressed by the acid dissociation constant, or K_a:

$$K_a = \frac{[H+][A-]}{[]}$$

The pK_a is the negative logarithm of the K_a. Strong acids have small pK_as. Looking at the above equation, it can be seen that when $[A^-] = [HA]$, then $K_a = [H^+]$. Then the $pK_a = pH$. This is the basis for the Henderson-Hasselbalch equation:

$$pH = pK_a + \log\frac{[H+][A-]}{[HA]}$$

This is Henderson-Hasselbalch Equation.

TITRATION CURVES

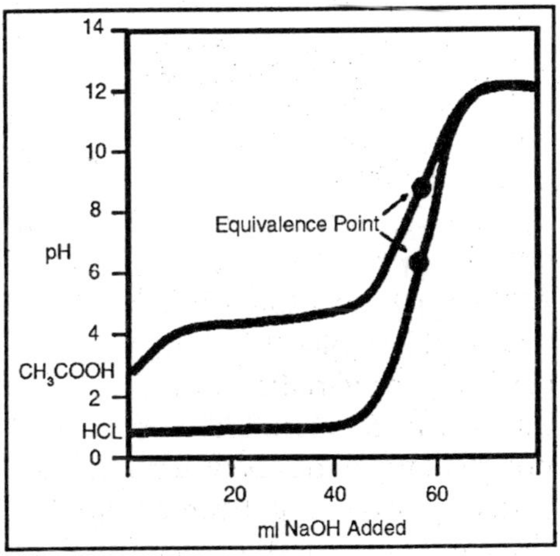

Fig. Titration Curves of Both a Strong and Weak Acid with a Strong Base.

When acids or base are added to a solution, the pH changes. The controlled addition of an acid or base to a solution is called a titration. A titration curve can be made by plotting the pH changes against the volume of acid or base added to a solution. Many titration curves are made by plotting pH on the y axis and the volume of base added to the solution on the x axis. For example, pH changes to acetic acid, CH_3COOH, and hydrochloric acid, HCl, with 0.1M sodium hydroxide, NaOH, is shown to the left.

Notice the shape of the curves. The CH_3COOH curve starts out at a higher pH than the HCl curve. This is because CH_3COOH is a weaker acid than HCl. As NaOH is added to the solution of CH_3COOH, the pH rises. However, the addition of NaOH to the HCl solution does not significantly change the pH. It takes less NaOH to change the pH of a weak acid solution than it does to change the pH of a strong acid solution. Eventually all of the acid in both solutions will have reacted with base, such that adding more base quickly increases the pH. This is called the equivalence point: This is the point where all the acid has reacted with the base, such that further additions of base quickly raise the pH. The equivalence point for a strong acid-strong base titration is pH = 7. For weak acids-strong base titrations, the equivalence point is pH >7.

BUFFERS

As shown above, the pH of a solution is dependent on the concentration of H^+ ions. Addition or removal of H^+ ions, then, can greatly affect the pH of a solution. In the body, the pH of cells and extracellular fluids can vary from pH 8 in pancreatic fluid to pH 1 in stomach acids.

The average pH of blood is 7.4, and of cells is 7 – 7.3. Although there is great variation in pH between the fluids in the body, there is little variation within each system. For example, blood pH only varies between 7.35 – 7.45 in a healthy individual. Large changes in pH can be life threatening. How does the body maintain a constant blood pH? The body uses a buffer system to withstand changes in pH. Buffers are made up of a mixture of a weak acid with its conjugate base or a weak base with its conjugate acid. Remember that an acid donates a H^+. A weak acid does not donate its H^+ as easily. Similarly, a weak base will not accept a H^+ as well as a strong base.

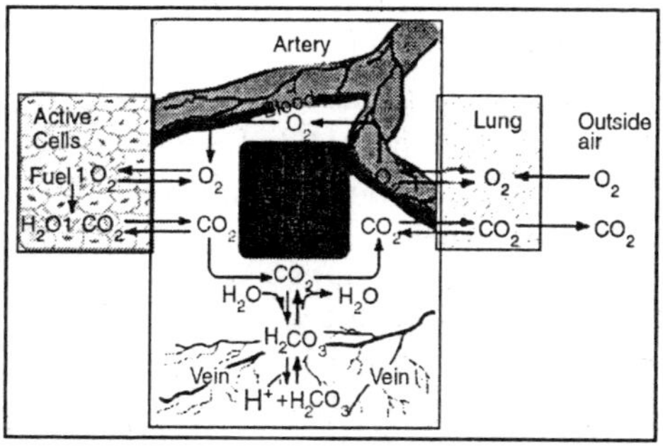

Fig. The Bicarbonate Blood Buffering System

Buffers maintain pH by binding H^+ or OH^- ions. This stabilizes changes in pH. The bicarbonate buffer system maintains blood pH near pH 7.4. The carbonic acid, H_2CO_3, in the blood is in equilibrium with the carbon dioxide (CO_2), in the air. Buffers are most effective in a pH range near its pK_a. This is where the titration curve is most shallow, and where the pH is least affected by added acid or base. For example, in the titration curve of phosphoric acid.

$$H_3PO_4 \overset{pK_a1}{\rightleftharpoons} H_2PO_4^- + H^+ \overset{pK_a2}{\rightleftharpoons} HPO_4^{-2} + H^+ \overset{pK_a3}{\rightleftharpoons} PO_4^{-3} + H^+$$

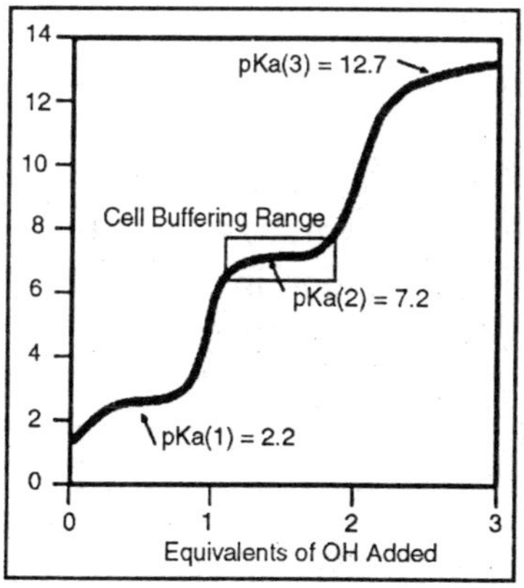

Fig. The Titration of a Polyprotic Weak Acid

The region in the rectangle is the buffering region of biological importance. Note that the slope of the curve is shallow within one pH unit of the pK_a value. This is where the buffering range is most effective, meaning that the buffer is able to resist changes in pH.

In chemistry, the degrees of acidity or alkalinity of a substance are expressed in pH values. The neutral point, where a solution would be neither acid nor alkaline is pH 7. Increasing acidity is expressed as a number less than 7, and increasing alkalinity is a number great than 7. Maximum acidity is pH 0 and maximum alkalinity is pH 14. Because each unit on the scale represents a logarithm, there is 10-fold difference between each unit (pH 5 is 10 times as acid as pH 6 and pH 4 is 100 times as acid as pH 6. Expressed mathematically, pH is logarithm of the hydrogen ion concentration divided into one - Log 1/H+

An acid is a substance that releases hydrogen into a solution. An alkaline substance is one that removes hydrogen from a solution. Most body functions occur at only certain levels of acidity or alkalinity, as well as, many enzymes and chemical reactions work best at a specific pH. A minute change in pH can have a profound effect on body function.*

Some of the pH values of a number of biological materials are given below:

Table. A - pH Values of Representative Biological Materials

Material	pH value
Blood, normal limits	7.3 - 7.5
Blood, extreme limits	7.0 - 7.8
Enzymes, Activity Range of	
Amylopsin, optimum	7.0
Erepsin, optimum	7.8
Invertase, optimum	5.5
Lipase, optimum	7.0 - 8.0
Maltase, optimum	6.1 - 6.8
Pepsin, optimum	1.5 - 2.4
Trypsin, optimum	8.0 - 9.0

Fruit Juices	
Apple	3.8
Banana	4.6
Grapefruit	3.0 - 3.3
Orange	3.1 -4.1
Tomato	4.2
Gastric juice (adult)	0.9 - 1.6
Milk (cows)	6.2 - 7.3
Plants (Extracted Juice)	
Alfalfa tops	5.9
Carrot	5.2
Cucumber	5.2
Peas, field	6.8
Potato	6.1
Rhubarb, stalks	3.4
String beans	5.2
Sweat	4.5 - 7.1
Saliva	6.2 - 7.6
Urine (human)	4.2 - 8.0
Tears	7.2

pH - POTENTIAL OF HYDROGEN DESCRIPTION

What Is Hydrogen

Hvdrogen (H) is the lightest and simplest element. A Hydrogen atom is made up of one proton and one electron. At room temperature, Hydrogen is gas - as it cools it becomes liquid. Hydrogen is one of 9 essential macronutrients and is a major component of organic molecules. Hydrogen supplies building blocks of all organisms whether plant, animal or human. Each water molecule contains two Hydrogen molecules and one Oxygen molecules (H20).

Ideal pH V.alues

- *When pH* = 8: There are 6 billion H+ ions in solution.
- *When pH* = 7: There are 60 billion H+ ions in solution.
- *When pH* = 6: There is 600 billion H+ ions in solution.

Ideal pH Values Of Blood, Urine and Saliva

- Arterial Blood – 7.40 – 7.45
- Capillary – 7.35 – 7.45
- Venous – 7.30 – 7.35
- Saliva – 6.5
- Urine – 6.8

pH Values Of Body

Saliva – (6.0 – 7.4)	Skeletal Muscle – (6.9 – 7.2)
Gastric Secretion – (1.0 – 3.5)	Heart – (7.0 – 7.4)
Pancreatic Secretion – (8.0 – 8.3)	Liver – (7.2)
Bile – (7.8)	Brain (7.1)
Small Intestine Secretion – (7.5 – 8.0)	
Urine – (4.5 – 8.0)	
Arterial Blood – (7.4 – 7.45)	
Capillary Blood – (7.35 – 7.4)	
Venous Blood – (7.3 – 7.35)	
Feces – (4.6 – 8.4)	

CLINICAL ACIDOSIS AND ALKALOSIS

The blood pH is supposed to remain "practically constant" according to our best text books. The physician who starts to check his patients, however, cannot occur. Instead of the normal figure of pH 7.3 to 7.5 he finds it varying from 7.2 to 7.8 (1).

The measurement of pH is attended with difficulty because the blood loses carbon dioxide (CO_2) and thereby rises the pH rapidly after it has been withdrawn. In two minutes a rise of.2 may occur (1). pH meters are available where the electrodes are built into a syringe so the pH may be read directly before the blood is exposed to air.

The importance of clinical pH testing was illustrated by the following report of a pharmacist. He had been supplying for a number of months a capsule formula of aspirin, caffeine, phenacetin and ammonium chloride to an arthritic patent where the remedy had been successful in controlling arthritic pains. The capsule dispensed was a standard formulas of a pharmaceutical house, with a specific catalog number. The patient returned two days after receiving a renewed supply of capsules, complaining that they were ineffective, and different from the previously used kind. On checking the records, it was found that the original formulas had been revised, and the supposedly unnecessary ammonium chloride had been omitted. On supplying capsules of ammonium chloride alone, the patient reported the usual successful results.

What was the biochemistry behind this: Simply that a high pH tends to throw calcium out of solution in the body fluids, and bursitis, arthritis, neuritis, lumbago, sciatica, and a host of other painful syndromes develop as a consequence. Cheilosis (lips redden, fissures at the angles) may occur and herpes simplex, and other virus types of diseases become active. The low calcium bicarbonate in the body fluids seems responsible, Vitamin F and Vitamin D as well as the Vitamin C complex are also factors, their deficiency aggravates the situation. Allergies become acute, the calcium deficiency aspect of allergic sensitivity is well known. Normal rates of wound healing are greatly reduced, ulcers then too become static.*

Ammonium chloride is a natural constituent of gastric juice, calcium chloride is the form of calcium found in arrowroot starch, so both of these materials are physiological therapeutic agents, factors common to body biochemistry and not new and foreign substances to the physiological economy (as are synthetics and antibiotics).

It is described that Ammonium Chloride is useful in the treatment of bronchial affections, hepatic congestion, pelvic cellulitis, muscular rheumatism, gout, sciatica, chronic glandular enlargement, hemicrania, senile gangrene, dysmenorrhea, leucorrhea.

Calcium Chloride is listed as useful in the treatment of hemorrhage, hemorrhagic endometritis, menorrhagia, erytherma nodosum, tetanus, spasmophillia, blackwater fever, hay fever, asthma, hemophilia, albuminuria, nephritis, typhoid, coryza, tuberculosis, osteomalacia, scrofula, rachitis, arthritis, spasm of glottis, infant convulsions, urticaria, pruritis, among others. All of these effects are no doubt in the main simply accomplished by correcting the unbalanced state of the buffers and mineral salts in the tissue fluids. It will be noted that Vitamin C complex is a desirable synergist in most of the conditions listed. Gout is mentioned as one situation where ammonium chloride may help. Recently it has been discovered that the juice of cherries contains some active agent effective in gout, 4 to 6 ounces of the juice (preferably unsweetened, but may be canned or cooked form) promoting relief with in a few days with consistent improvement.

This is the first time that the possibility of gout being a deficiency disease has been suspected. We can say that the ingestion of sodium compounds particularly sodium phosphate - is definitely aggravative. Gout patients should see that their diet contains a preponderance of potassium. Sodium promotes the precipitation of gouty concretions of sodium urate in the tissues.

A very important aspect of alkalosis is the fact the habitual use of milk of magnesia promotes the condition. It also tends to cause epistaxis (nosebleed - can be secondary to nasal and sinus issues; scarlet fever or typhoid; arteriosclerosis; hypertension and anemias), very specifically. You will be surprised how many of the patients who ask for a remedy for nosebleed are using milk of magnesia as a laxative.

Citrus fruits being of a highly alkaline ash, with a high content of organic acid (citric), need special attention. The first and immediate effect of ingestion of a few ounces of grapefruit or lemon juice is to lower (acidity) the blood pH. The patient with alkalosis feels temporarily better. Later in the day, after the citric acid has been destroyed by oxidation as a fuel (it is classed as carbohydrate) the aggravation of the alkaline state become apparent. The temporary effect of the citric acid is to cause calcium to be picked up - no doubt from some bone reserves - and after the pH change this calcium is deposited elsewhere - in physiologically undesirable spots. The treatment for bursitis obviously becomes simple. No longer can we recommend X ray treatment with equanimity. Let the physiological treatment be the preferred method.

Guanidine, a fatigue and tissue poison, and end product of the breakdown of creatine (combated by Vitamin E complex which stops the loss of creatine from the tissues), is the most potent organic alkaline substance know. It specifically precipitates calcium (recall our comment that it is diffusion from tired heart muscle into coronary vessels is the cause of precipitation therein of calcium), is normally reconverted into creatine by the influence of the parathyroid hormones, and with the assistance of the thyroid. Blood guanidine levels rise eight fold after parathyroidectomy.

We can see where the guanidine effect as a fatigue poison is contributory to the arthritis, sciatica, etc., the patient often tell us how fatigue aggravates his state of misfortune. Local inflammation also can release guanidine, its presence insure a spastic state of blood vessels, contributing to gangrene, of necessity.

There seems no doubt that the effect of oxygen metabolizing vitamins - the Vitamin C and Vitamin E complexes - is vital in combating these morbid reactions by preventing the degradation of tissue elements into guanidine. Where Vitamin E deficiency has caused the tissue demand for oxygen to rise up to 250%, of normal, the augmented release of end products certainly includes guanidine, and we see here how the find of Dr. Shute of London, Ontario, that Vitamin E reduces capillary hemorrhage may be rationalized.

A vicious cycle is set up, the more oxygen demand the less supplied by reason of the constrictive effect of guanidine on blood vessels. Degeneration, rupture of capillaries is inevitable. As Vitamin E promotes more oxygen supply, that is why it also is good in gangrene (especially the diabetic type) and in "capillary fragility". (The true antifragility vitamin is Vitamin P, which provides a special calcium to promote collagen formation.)

Vitamin F opposes the toxic effect of guanidine by providing more of the diffusible calcium (from the colloidal blood reserves) that is precipitated out of the body fluids by guanidine, thereby ameliorating the spastic, nervous and irritative (allergic) reactions.

The thyroid hormone physiologically promotes the resorption and dissolution of protein structures that have reached the end of their physiological cycle (the theory of the dynamic state of living tissue, replaced and rebuilt at specific time intervals). That is why in children, thyroid deficiency is the cause of delayed development. In the adult, a hyperactive thyroid may again cause toxicosis by promoting more tissue poisons than the eliminative system can tolerate. In old people, this may be a critical situation, thyroid sometimes becoming a violent poison.

Only because the normal synergists are not available to maintain the desirable complete cycle of activity. We can see that Vitamin F is vital to the prevention of this toxicosis. It tells why the thyroid puts its secretion into the blood after ingestion of Vitamin F to the extent of a doubling of the iodine content of the blood. The normal activity of the gland had been blocked by reason of Vitamin F deficiency. In this blocking, we have apparently the explanation for prostate hypertrophy. Once the thyroid is permitted to secrete, its hormone makes quick work of the fibrous tissue collected in the prostate. This also explains why we consider the Vitamin F complex, and Vitamin F2 the most friendly vitamins for the older person. The Vitamin F2 is a more highly developed compound of the basic F (a sensitized fatty acid), in which this fatty acid is combined into a phospholipid molecule that appears to have a specific function of catalyzing protective (Insulating) layers in nerve tissue and cell nuclear structure. Thereby protecting the basic controls of metabolic activity. Loss of appetite and wasting disease suggests the possible need for this vitamin. Appetite restoration is immediate where the deficiency exists.

There seems to be as much clinical acidosis as alkalosis. The effects of acidosis may be aggravated by a diabetic state where organic acids cannot be oxidized normally, instead must be combined with reserve alkaline salts and excreted. Where sodium bicarbonate is administered to relieve diabetic acidosis, the urine should be watched and dosage stopped when a neutral is reached, according to this authority. Normally, blood glutamine provides ammonia for this purpose. Beet root and leaf are best vegetable sources of glutamine, is very high if grown on high ammonia content soils. (Plant cells form glutamine to dispose of excess ammonia).

When abnormal blood pH is found, it must be considered only as a sign that something is wrong. The symptoms may be entirely different with the same abnormal pH figure in different patients, in view of the fact that there is an infinite number of combination of chemical situations that could alter pH - as many as there are different acids and alkalies. Just as a plump patient may be bloated with gas, water logged, fat, or hyper muscular. It is a preliminary classification, not a final.

The pH of the blood is a resultant of the forces acting upon it. The regulation of the blood CO_2 (carbon dioxide) by respiration control Centre of the brain is a major factor. The parathyroid by eliminating the alkaline guanidine, and the kidney, eliminating mineral salt (both acid and alkaline as may be required) no doubt exercises a supplementary control.

The clinical value of vinegar is more or less well known as an alternative of merit. Creatine is methylguanidine acetic acid, and is often craved by the patient, very logically it would seem. Cider vinegar being the preferred form, no doubt the malic acid of the apple is a factor, calcium malate being a catalyzer of polyphenol detoxification.

ACIDOSIS AND ALKALOSIS INFORMATION

Regulatory Factors

The following factors tend to promote normalization of pH through physiological mechanisms. They do not supply mineral elements which directly influence body chemistry. To directly influence pH values through acid or alkaline mineral administration, see: Alkalosis or Acidosis. Normalization of the function of the glands and kidneys which regulates pH balance should be the primary consideration. However, in most cases specific support of either acid or alkaline mineral therapy is necessary where immediate results are to be obtained, particularly in acute situations.

Primary

- *Renatrophin PMG*: Kidneys are thought to be the most important organ in acid-base balance regulation.
- *Drenatrophin PMG*: Adrenal support regulation of sodium-potassium-chloride balance.
- *Thytrophin PMG*: Thyroid supports regulation of calcium-magnesium-phosphorus balance.

Normally the endocrine glands, along with the kidneys regulate the pH of the blood, just as the pancreas with it insulin regulated the blood sugar levels. The sex glands, along with the adrenals, seem to be the main endocrines involved. Particularly after menopause or male climacteric there may be serious changes in the body's economy whereby acid-base disorders develop, which may be aggravated, if not caused by dietary circumstance - just as diabetes is aggravated by a high sugar intake.*

Synergistic Supplements

- *Pituitrophin PMG: Pituitary is the hormonal control of* endocrine system
- *Ovatrophin PMG*: Specific cell activator
- *Utrophin PMG*: Specific cell activator
- *Prostate PMG*: Specific cell activator
- *Orchic PMG*: Specific cell activator

ALKALOSIS

PHYSIOLOGICAL CONSIDERATIONS

The calcium may precipitate out of body fluids (with excess tissue calcium) where alkalosis exists. Increased contractility of muscles noted. There may be calcium deficiency symptoms, but cannot assimilate calcium in any form. The neuritic pains which may develop are no doubt due to calcium crystals (calcium carbonate) that form at nerve endings which accumulate and disperse according to the variations of the pH of the blood that follow dietary levels of acid and alkaline food ingestion, unable to be compensated for by a depleted endocrine system.

FREQUENTLY OBSERVED ENTITIES

Menopausal Disorders; Allergic States; Osteoarthritis; Renal Calculi; Prostatitis; Bursitis; Bronchitis; Asthma; Chronic indigestion; Coronary Sclerosis; Hypercholesterolemia.

PREDISPOSING FACTORS

- Endocrine insufficiency (kidneys, thyroid, adrenals, gonads, etc)
- Excess carbohydrate intake (citrus fruits, sugar, etc.)
- Excessive alkaline ash foods
- Mental Stress
- Fluid loss (perspiration, diarrhea, vomiting, etc)

Primary

- *Cal-Amo*: Chloride ion source.
- *Phosfood Liquid*: Orthophosphoric acid source.
- *Soy Bean Lecithin*: Phosphorus and choline source.
- *Betaine Hydrochloride*: Hydrochloric acid source.

Secondary

- *Cataplex V G*: Acetic acid metabolism
- *Bio-Dent*: Bone-source of phosphorus
- *Calcifood*: Bone source of phosphorus
- *Biost*: Phosphatase source (enzyme)
- *Prost-X*: Phosphatase source (enzyme)

METABOLIC ALKALOSIS

The pH scale runs from 0-14 with 7.0 be a neutral pH and with a pH of 6.4 to 6.8 being the most suggested when considering your biological terrain first thing in the morning (urine and saliva)- it will change to alkaline numbers throughout the day with 7.0 - 7.8 being the norm a couple hours after eating.

Causes Of Metabolic Alkalosis

All diuretic therapies except for the carbonic anhydrase inhibitors. These diuretics cause a tremendous increase in the concentration of Na+ ions brought into the distal and collecting tubules. This lead to rapid reabsorption of these sodium ions and creates an enhance hydrogen ion secretion due to the Na+-H+ exchange mechanism. Na+ get reabsorbed and H+ get suddenly excreted. Excessive ingestion of alkaline drugs (Rx of gastritis and peptic ulcers)

WATER, PH, AND NON-COVALENT BONDING

WATER

Water is essential for life. It covers 2/3 of the earth's surface and every living thing is dependent upon it. The human body is comprised of over 70% water, and it is a major component of many bodily fluids including blood, urine, and saliva. What accounts for the ubiquitous use of water in living systems? If we take a step back and consider the structure of water and compare it to another substance, methane, we can understand the unique properties water possesses that make it well suited for biological systems.

Water (H_2O) is made up of 2 hydrogen atoms and one oxygen atom, with a total atomic weight of 18 daltons. The structure of the electrons surrounding water is tetrahedral, resembling a pyramid. For comparison, Methane (CH_4) is made up of one carbon and 4 hydrogens. Note that methane is similar to water in that it weighs 16 daltons and also has a tetrahedral structure, yet has very different physical properties.

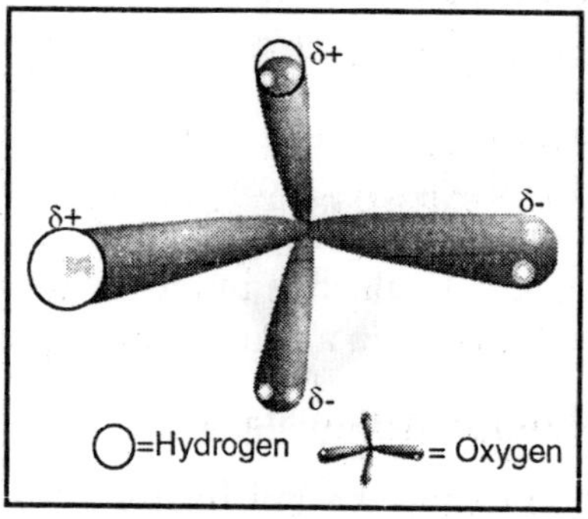

Fig. Molecular Geometry of Water: The Orbitals are Pyramidal, Yet the Atoms form a Bent Shape.

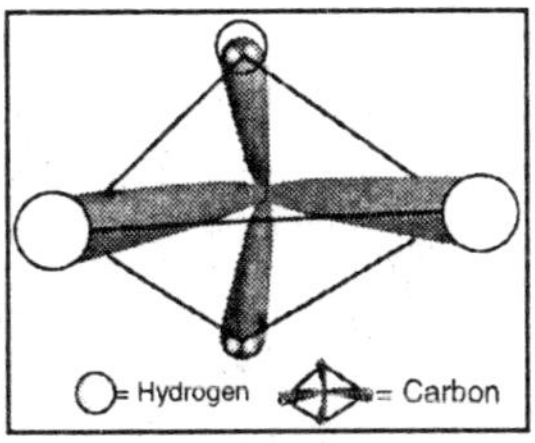

Fig. Molecular Geometry of Methane: Pyramidal

All the vertices on the methane that form the points of a pyramid are occupied by a hydrogen. However, only two vertices in water are occupied with hydrogens. The other two vertices are each occupied by a lone pair of electrons. This fact causes the water molecule to have a bent molecular shape. Oxygen is highly electrophilic (electron loving). This means that even though the oxygen in water is bound to each of the hydrogens by a covalent bond (sharing a pair of electrons), the oxygen "pulls" the shared electrons closer to itself. This unequal sharing of the electrons in the O-H bond in water causes the hydrogens to have a partial positive charge (positive dipole), and the oxygen has a partial negative charge (negative dipole). Water is called a polar molecule because it has a positive side and a negative side, called a dipole moment. In contrast, the carbon in methane shares electrons equally with the 4 hydrogens. Methane's does not have a dipole moment, and in contrast with water is non-polar.

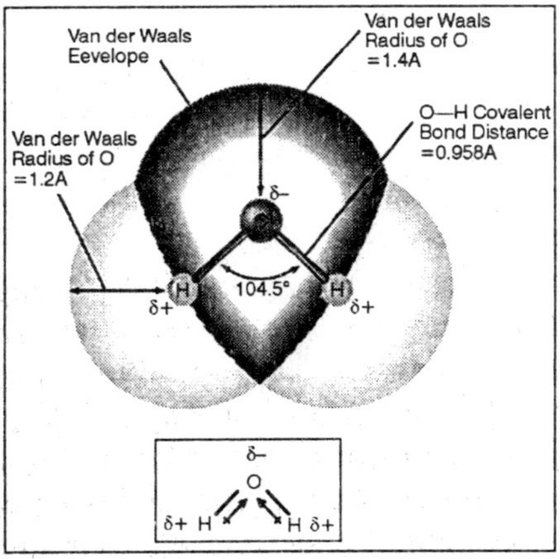

Fig. The "Magnet-like" Dipole Moment in Water

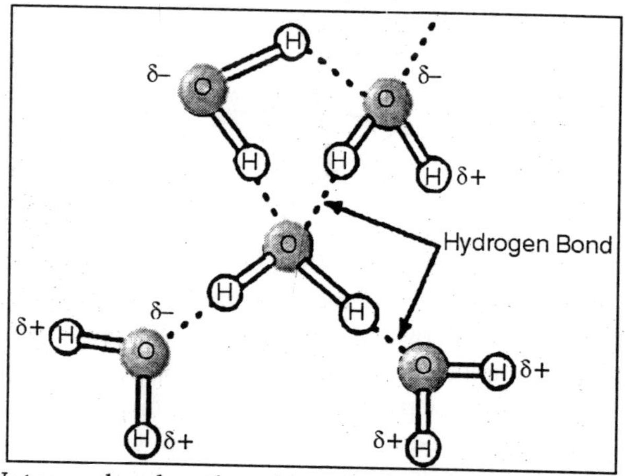

Water molecules, then, have both a partial positive and a partial negative charge. Since opposites attract, this means that water molecules are attracted to each other, like the positive and negative ends of a magnet. Because of these attractive forces, water molecules are in close proximity to one another, making water very dense. Methane molecules do not have a dipole attraction for one another, and therefore and are spaced farther apart.

Therefore, despite its similar size and mass to water, methane is much less dense than water. This is the reason that under room temperature situations, water exists as a liquid while methane is a gas. We know that water can be converted to its gaseous phase, steam, but only by applying a lot of energy in the form of heat to disrupt the large attraction of the water molecules for one another.

Hydrogen Bonding

Water molecules are bound together through hydrogen bonds. Hydrogen bonds arise when a hydrogen that is covalently bound to either an oxygen, sulfur, or nitrogen atom nears an electrophilic atom, such as oxygen. The electrophilic atom "pulls" the hydrogen closer to itself. The end result is that the hydrogen is now shared (unequally) between the atom to which it is covalently bound and the electrophilic atom to which it is attracted (O-H...O). Since water is made up solely of hydrogens covalently bound to oxygens and electrophilic oxygens, it is easy to see why there are so many hydrogen bonds in water. The molecules in water are in constant motion and the hydrogen bonds are constantly being broken and reformed. In fact, the average hydrogen bond lasts only a one-trillionth of a second (10^{-12} s)!

Each hydrogen bond has an average energy of 20 kJ/mol. This is much less than an O-H covalent bond, which is 460 kJ/mol. Even though an individual hydrogen bond is relatively weak, the large number of hydrogen bonds that exist in water which pull the molecules together give water its special density and phase transition properties. This is much like the story of Gulliver's Travels, when Gulliver awoke on the island of Lilliput to find himself tied down with many tiny strings. Each string was individually weak, but together they were enough to bind Gulliver.

On average, a liquid water molecule will have 3 out of 4 possible hydrogen bonds. This is because the molecules in liquid are in constant motion. In ice, water will form all 4 possible hydrogen bonds because the molecules in a solid are essentially locked in place. Water molecules are further apart from one another in ice than they are in the liquid state. Ice is therefore not as dense as liquid water.

This is why icebergs float in the ocean and why ice floats in a beverage. This contrasts greatly from the normal relationship between a compound's solid and liquid forms, with the solid form usually being of a higher density. While the unusual phase density properties of water might not seem too important in biological reactions, it is very important for life on earth. If ice did not float, then in very cold bodies of water ice would sink, removing its insulating effects from the water's surface. Lakes, rivers, and major portions of the oceans thus would freeze completely from the bottom up. The world would be a much different place, and much less suited for life.

Water is a Good Solvent

The polar nature of water, with its partial positive and partial negative dipole, allows it to dissolve charged molecules (ions) easily. Water is thus an excellent solvent for charged compounds. The positive side of water surrounds negatively charged molecules, and the negatively charged side of water surrounds positively charged molecules.

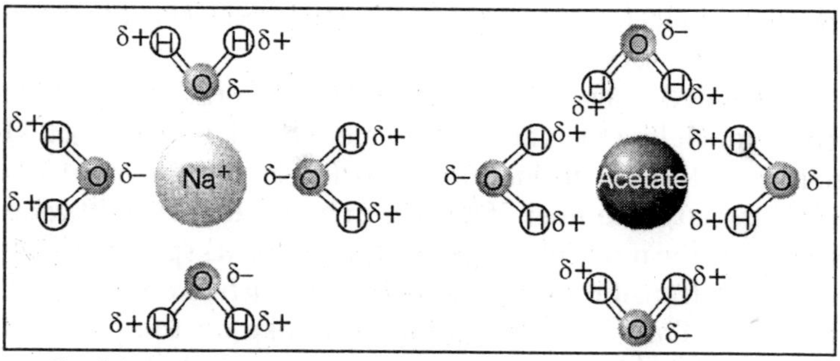

In this way, water makes "solvation shells" around ions. Water can also readily dissolve other polar molecules, even if they are not positively or negatively charged. The solvent properties of water allow for dissolved metals and buffering systems that are very important for the workhorses of life, enzymes. However, the saying "oil and water don't mix" is true—water cannot dissolve oil. This is because oily substances are non-polar. Non-polar substances (which lack dipoles) are also called *hydrophobic* (water fearing). Hydrophobic substances gather together to exclude water as best they can. This is why you see oil droplets in water. This is also important for the stability and structure of enzymes.

NON-COVALENT BONDING

Non-covalent bonds are not at strong as covalent bonds, but they are important in the stabilization of molecules. In contrast to covalent bonds, non-covalent bonds do not share electrons.

Noncovalent bonds include:

- Electrostatic interactions
- van der Waals forces
- Hydrophobic interactions
- Hydrogen bonds

Electrostatic interactions are formed between positive and negative ions. The bond is non-directional, meaning that the pull of the electrons does not favour one atom over another. An example is NaCl, which is formed between the positively charged Na^+ ion and the negatively charged Cl^- ion. The average strength of an electrostatic interaction is 15 kilojoules/mol (kJ/mol). The bond strength lessens when the distance between the two ions increases.

Van der Waals forces are weak forces between temporary dipoles. These forces may be attractive or repulsive. They are also non-directional. The average bond energy for van der Waals forces is on the order of 10 kJ/mol. Hydrophobic interactions result when non-polar molecules are in a polar solvent, *e.g.* H_2O. The non-polar molecules group together to exclude water (hydrophobic means water fearing). By doing so they minimize the surface area in contact with the polar solvent.

A hydrogen bond results when a hydrogen atom that is covalently bound to an electronegative atom (*e.g.* O, N, S) is shared with another electronegative atom. A hydrogen bond is directional towards the electronegative atom. An example of this is the hydrogen bonds formed in water. Hydrogen bonds are constantly being made and remade. The half-life of a hydrogen bond is about 10 seconds. A hydrogen bond has an energy of 21 kJ/mol. As stated above, non-covalent bonds are not as strong as covalent bonds, but the additive effect of many non-covalent bonds can stabilize a molecule. Non-covalent bonds are very important in the structure of proteins.

Bond type	**Energy (kJ/mol)**
Covalent, *e.g.* C-C	350
Electrostatic	15
Van der Waal's	10
Hydrogen	21

TEST AND PROBLEMS

pH Questions

Below are a few sample questions to elucidate the concepts discussed in the book:

- A simple pH problem

- pH of a weak acid
- A buffer problem
- A more involved buffer problem

A Simple pH problem

What is the pH 0.1 M HCl?

Answer: The pH of a solution is the negative logarithm of the concentration of H^+ ions. In this case, HCl is a very strong acid. HCl completely dissociates to H^+ and Cl^-.

$$HCl \rightarrow H^+ + Cl^-$$

The concentration of HCl is 0.1 M, if all of it dissociates to H^+, then

$[H^+] = 0.1$

$-\log(0.1) = 1.26$

$pH = 1.26$

Remember that a pH of 7 is neutral, and that the pH scale is logarithmic, meaning each pH unit represents a factor of 10. This HCl solution has a difference of 5.74 pH units from water (7 - 1.26). Taking the inverse log of 5.74 means that this HCl solution is more than 500,000 times as acidic as water.

pH of a Weak Acid Solution

What is the pH of a 0.05 M solution of acetic acid? (The K_a of acetic acid = 1.8×10^{-5})

Answer: Determining the pH of a weak acid solution is more complicated than for a strong acid. This is because a weak acid does not completely dissociate.

$$CH_3COOH \rightleftharpoons H^+ + CH_3COO^-$$

To simplify matters, it is helpful to construct a grid:

Reaction	**CH_3COOH**	**H^+**	**CH_3COO^-**
Initial	0.05 M	0 M	0 M
Change	(–x) M	+ x M	+x M
Equilibrium	(0.05–x) M	x M	x M

To determine the $[H^+]$, we must employ the K_a of acetic acid, which is 1.8×10^{-5}.

$$Ka = \frac{[H+][CH_3COO-]}{[CH_3COOH]} = 1.8 \times 10{-}5$$

Substituting the equilibrium values in the table above, we get:

$$1.8 \times 10\text{–}5 = \frac{(x\ M)(x\ M)}{(0.05 - x)\ M}$$

Since acetic acid is a weak acid, we can assume that very little of it will dissociate. This simplifies the above equation to:

$$1.8 \times 10^{-5} = \frac{(x\ M)(x\ M)}{0.05}$$

This can be easily solved:

$9 \times 10^{-7}\ M = x^2\ M$

$x = 9.5 \times 10^{-4}\ M$

$x = 0.00095\ M$

Now that we have the [H+], we can solve for pH:

pH = –log[0.00095]

pH = 3

Buffer Problem

Question 1: What is the ratio of [lactate]/[lactic acid] at pH = 5? (The pK_a of lactic acid = 3.85)

Answer: To solve this problem, we need to utilize the Henderson-Hasselbalch equation:

$$pH = pKa + \log \frac{[H+][A-]}{[HA]}$$

Now just substitute into the equation:

$$5 = 3.85 + \log \frac{[\text{lactate}]}{[\text{lactic acid}]}$$

$$1.15 = \log \frac{[\text{lactate}]}{[\text{lactic acid}]}$$

$$14 = \frac{[\text{lactate}]}{[\text{lactic acid}]}$$

There is 14 times as much lactate as lactic acid at pH 5.

Question 2: What happens if 2 ml of 100 mM HCl is added to 100 ml of a 10 mM solution of phosphoric acid at pH = 7?

Answer: This is a more involved problem. The first thing to Realise is that phosphoric acid is polyprotic, meaning it has more than one H+ to donate, and therefore more than one pKa.

$$\begin{array}{ccc} pK_a1 & pK_a2 & pK_a3 \end{array}$$

$$H_3PO_4 \rightleftharpoons H_2PO_4^- + H^+ \rightleftharpoons HPO_4^{-2} + H^+ \rightleftharpoons PO_4^{-3} + H^+$$

Since this problem is at pH = 7, it is the pKa2 value that is important, for this is the pKa nearest the pH.

First we need to find the ratio of,

$$[HPO_4^{-2}]/[H_2PO_4-]:$$

Again, use the Henderson-Hasselbalch equation:

$7 = 7.2 + \log[HPO_4^{-2}]/[H_2PO_4^-]$

$0.63 = [HPO_4^{-2}]/[H_2PO^{4-}]$

$0.63[H_2PO^{4-}] = [HPO_4^{-2}]$

We know we have a total of

(0.01 moles/L) × (0.1 L) = 0.001 moles phosphoric acid that at pH = 7 is made up of HPO_4^{-2} and H_2PO_{4-}. (Note 10 mM = 0.01 moles/L and 100 mL = 0.1 L)

0.001 moles = 1 mmole = $HPO_4^{-2} + H_2PO^{4-}$

We know from above that $[HPO_4^{-2}] = 0.63[H_2PO^{4-}]$

so 1 mmole=$0.63[H_2PO_4-] + [H_2PO_4-] = 1.63[H_2PO_4-]$

$[H_2PO_4-] = 0.6$ mmoles

and so then $[HPO_4^{-2}] = 0.4$ mmoles

So far, this problem is much like the previous one. Now let's answer the question about the addition of 1 ml of 0.1 M HCl. From the high pK_a2 value, we know that phosphoric acid is a weak acid. It will react completely with the HCl (which is the same as H+ in solution):

pK_a3

$HPO_4^{2-} + H^+ \rightleftharpoons H_2PO_4-$

By adding 2 ml of 0.1 M HCl, we are adding:

$(1 \times 10^{-3}$ L) × (100 mmoles/L) = 0.2 mmoles H^+

If we set up the grid as shown above,

Reaction	HPO4-2	H+	H2PO4-
Initial	0.6 mmoles	0.2 mmoles	0.4 M
Change	–0.2 mmoles	–0.2 mmoles	+ 0.2 M
Change after equilibrium	0.4 mmoles	0 mmoles	0.6 M

What is the pH of the phosphoric acid solution after the addition of HCl?

Again, use the Henderson-Hasselbalch equation to solve,
pH = 7.2 + log(0.6/0.4)
pH = 7.37

Note that this is not a big change. This is because of phosphoric acid is a buffer near its pKa and as such is able to resist changes in pH.

BUFFERING OF WATER

A solution that contains a weak acid and its conjugate base has the special property of being able to resist changing its pH when either a base or an acid is added to it. Such a solution is called a buffer because the solution is protected, or buffered, from pH changes even when H_3O^+ or OH- ions are added to the solution. Buffers help biochemists study biomolecular reactions in the Labouratory by stabilizing the pH of solutions used for experiments.

Fluids inside biological organisms must be maintained at a constant pH for the health of the organism. That is because these fluids, whether blood, cytosol, lymph, bile, etc., contain enzymes that carry out the reactions that support life, and each of these enzymes only has a certain pH range within which it can function effectively. This is called as the pH dependency of the enzyme.

If the pH is too high or too low, the amino acid residues important for catalysis might be in the wrong protonated state for catalysis, or the enzyme's conformation could change, rendering it nonfunctional. Shown at right is a graphic for a fictional enzyme that is very similar to the pH dependence of trypsin. Other enzymes in the same organism might have a very different dependence. For example, pepsin has optimal activity at a pH of 2, because it is a digestive enzyme and has to be active in acidic digestive fluids.

If the pH of biological fluids is so important, how does an organism make sure its fluids are all kept at the proper pH for the correct functioning of all of its enzymes, even if the organism ingests material that is acidic or basic? The answer is that all the biological fluids are buffered. Buffers are solutions that are resistant to changes in pH, even if acid or base is added to the solution. Because enzymes are so pH-sensitive, not only for their activity but also for their stability in solution, buffers are very important for biochemical research. In a lab, researchers keep all of their proteins in buffered solutions. Therefore, beginning biochemists need to understand how buffers work in the body and how to make them in the lab.

This difficult concepts section is a primer to help you grasp buffers to a level of understanding needed for introductory biochemistry. Go through the subsections in order to get the best instruction: In chemical and biological processes the pH value is an essential parameter. Slight changes in pH cause negative results between the reaction partners.

It is therefore necessary to maintain a correct pH and to protect the system from internal and external chemical influences. This work is performed by buffer substances, which assure a pH in a specific range.

Dissociation constants, relationship between pH, pKa and the buffering capability are mainly explained in the Henderson-Hasselbach-Equitation. Buffers have a characteristic buffer capacity, buffering range and dissociation constants at defined temperatures. Temperature shows effects on the pH of the buffer.

These differences could be quite small but could still cause significant discrepancies in biological systems. Good's Buffers are buffering substances which were selected and described by Norman Good and his collegues in 1966.

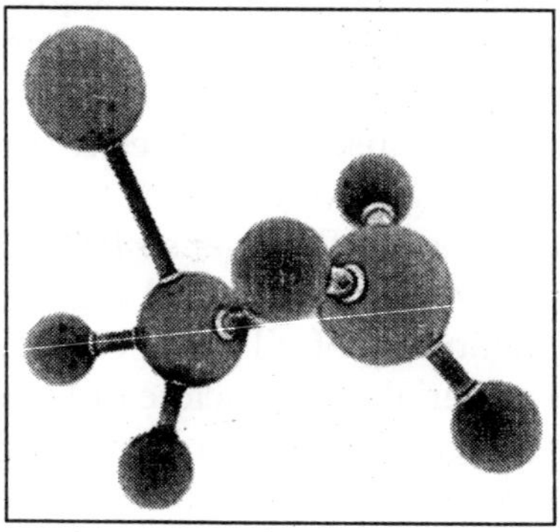

Sinus offers a selected range of buffers, which are widely used for research in biochemistry and biotechnology.

EXAMPLE

If 10 mL of 1M NaOH are added to one liter of a buffer that is 0.3 M acetic acid and 0.2 M sodium acetate ($Na+CH_3COO^-$), how much does the pH change? The pK_a for acetic acid is 4.76.

Answer: In the original solution, acetic acid is the weak acid, and acetate is the conjugate base. Thus, [base] = 0.2 M, and [acid] = 0.3 M.

Using the Henderson-Hasselbalch equation:

$$^{P}_{H} = pK_a + \log \frac{[Base]}{[Acid]}$$

pH = 4.72 + log (0.2/0.3)

pH = 4.54

NaOH is a strong base, and dissociates completely. Therefore, adding the strong base results in (0.01 liter)(1 M) = 0.01 moles of OH- ions into the solution.

All of the OH– ions will react with the acetic acid to form acetate ions:

$$CH_3COOH + OH- \rightleftharpoons CH_3COO- + H_2O$$

Thus, when NaOH is added, the number of acetic acid molecules in the solution will decrease, but the number of acetate molecules in the solution will increase. In this case, 0.01 moles of acetic acid will be used up to neutralize the 0.01 moles of OH-, and this will form 0.01 moles of acetate ion in the solution.

At the end of the reaction:

Moles of acetic acid	=	**Moles of acetic acid before the reaction**	–	**Moles of acetic acid used up in the reaction**		
	=	0.3 moles	–	0.01 moles	=	0.29 moles
Moles of acetate	=	moles of acetate before the reaction	+	moles of acetate made in the reaction		
	=	0.2 moles	+	0.01 moles	=	0.21 moles

Now we are almost ready to plug back into the Henderson-Hasselbalch equation to find the new pH of the solution. We just need to remember that the volume has changed slightly, from 1 liter (1000 mL) to 1 liter plus 10 mL (1010 mL). The Henderson-Hasselbalch equation requires that the [Base] and [Acid] be molar (M, or mol/L) quantities. However, notice how that because both the acid and base species are in the same volume, the volume correction cancels out:

$$pH = pK_a + \log\frac{[Base]}{[Acid]}$$

pH = 4.72 + log 1.38

pH = 4.86

The pH change is 4.86 – 4.54 = 0.32 units.

Example:

How much does the pH change when 10 mL of 1 M NaOH is added to 1 liter of water?

Answer:

Pure water has a pH of 7.0. NaOH is a strong base, and once it has been added to the water it will dissociate completely, contributing 0.01 moles OH- ions, in a total volume of 1010 mL (1 liter plus the 10 mL NaOH added). Thus,

$[OH^-]$ = 0.01 moles/ 1.01 liter = 0.0099 M

Plugging into the Kw expression: $[OH^-][H3O+] = 10–14$:

$(0.0099)[H_3O+] = 10–14$

$[H_3O+] = 1.01 \times 10–12$ M

Then plugging into the pH equation:

$pH = -\log [H^+]$

$pH = -\log 1.01 \times 10^{-12}$

pH = 11.99

The pH has shifted from the neutral pH of 7 for the pure water to a pH of almost 12 with the NaOH added, a shift of 5 pH units!

STABILIZING PH OF AQUEOUS SOLUTIONS

In example, the pH of the buffer solution changed by less than 0.5 units when the NaOH was added. Yet adding the same amount of NaOH to pure water resulted in a 5 unit pH shift in example 6. Clearly, the conjugate acid/base pair was able to protect the solution from a large change in pH. A buffer is a solution that resists a change in pH when acids or bases are added to it. How is this possible? Buffers work by acting a little bit like a sponge, soaking up excess H3O+ or OH– ions when they are added to a solution. The dissociation reaction for acetic acid shows that a solution of acetic acid contains acetate ions:

$$CH_3COOH + H_2O \rightleftharpoons CH_3COO- + H_3O^+$$

How do acetic acid and the acetate ion work to buffer the solution? If a strong acid, such as hydrochloric acid, is added to this buffer, the H3O+ ions generated will react with the acetate ions, removing them from the solution:

$$CH_3COO^- + H_3O+ \rightleftharpoons CH_3COOH + H_2O$$

acetate ion	hydronium ion	acetic acid	water

Similarly, if a strong base, generating lots of OH^- ions, were added to the acetic acid buffer solution

$$CH_3COOH+ HO- \rightleftharpoons CH_3COO - H_2O$$

acetate acid	hydroxide ion	acetate acid	water

The OH- ions are removed via a reaction with acetic acid. It is important to remember that buffers cannot maintain their pH indefinitely as more acid or base is added. Imagine slowly pouring a bucketful of water onto a small sponge on the kitchen floor. At first, as the water pours out of the bucket it is absorbed by the sponge, keeping the kitchen floor dry.

But once the sponge is soaked through, it can hold no more and the water spills onto the floor as fast as if the sponge weren't there. In the same way, buffers will protect the pH of the solution to some extent, but if they are inundated by large amounts of H_3O+ or OH– ions, all the available conjugate acid or base molecules will have been used up and pH changes will rapidly occur.

Thus, buffers only work as long as the amount of conjugate acid and base ions are large compared to the H_3O+amount of or OH– ions to be removed. Remember from the Henderson-Hasselbalch equation that the numbers of conjugate acid molecules is equal to the number of conjugate base molecules when the pH = pKa:

$pH = pK_a + \log([Base]/[Acid])$

if [Base] = [Acid], then

$pH = pK_a + \log 1$

$pH = pK_a$

Because the conjugate acid and conjugate base molecules are available in equal amounts when the pH = pKa, the buffer solution has the strongest ability to protect against pH changes caused by incoming H_3O^+ or OH-ions when the pH of the solution is close to the pK_a of the conjugate acid. If the pH of the solution strays too far from the pKa point, the buffer uses up all its available conjugate acid or base molecules, and the pH starts to fall dramatically. At this point, the buffering capacity of the buffer has been exceeded. this can be seen in the buffer titration graph below. Buffers protect the pH of a solution best within one pH unit of the pKa.

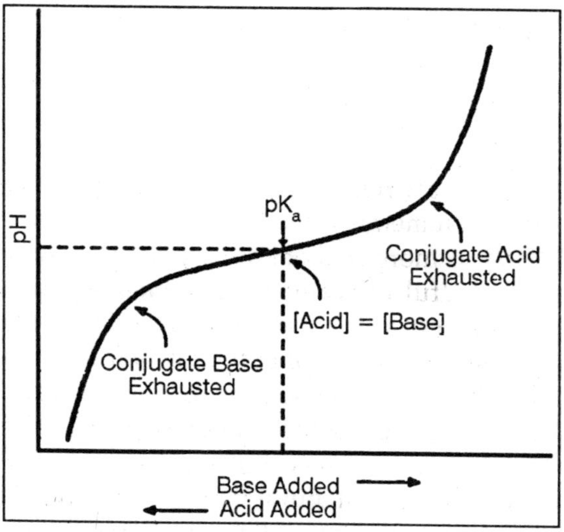

Fig. Buffer Titration: Within One Unit of the pKa, the Solution Resists Changes in pH.

Buffers are important for biochemistry because the structures (and therefore the functions) of biological molecules are stable within a relatively narrow range of pH values.The evolution of life, which is believed to have begun in water, was likely due in part to the stable pH of seawater. Even today, the largest buffered systems in the world are the Earth's oceans. Carbon dioxide (CO2) from the Earth's atmosphere reacts with H2O to produce carbonic acid, H2CO3. Carbonic acid in turn reacts with water to form bicarbonate:

$$H_3CO_3 + H_2O \rightleftharpoons HCO_3 - H_3O^+$$

crabonic acid water bicarbonate ion Hydronium ion

Furthermore, the bicarbonate ion can also react with water to form carbonate:

$$H_3CO_3^- + H_2O \rightleftharpoons HCO_3 - H_3O^+$$

bicarbonate ion water Carbonate ion hydronium ion

The carbonate and bicarbonate ions act together as conjugate acid/base pair to keep the pH of the ocean at about 8.2. In the living organisms, proteins (the molecules which do most of the"work" of a cell) are very sensitive to acid concentration, because they only fold into their proper three-dimensional shapes within a small pH range. Some of the individual amino acid building blocks that make up a protein have ionizable side chains that change their ionization state, or charge, at different pH. Thus, if the pH changes, some of the attractive forces that hold the protein together will also change. The ionization states of the side chain of the amino acid lysine are shown below. The pKa of lysine's side chain amino group is approximately 10.8. Therefore, at a pH lower than the pKa, the side chain is in the protonated (conjugate acid) state; whereas at a pH higher than the pKa, the side chain is unprotonated.

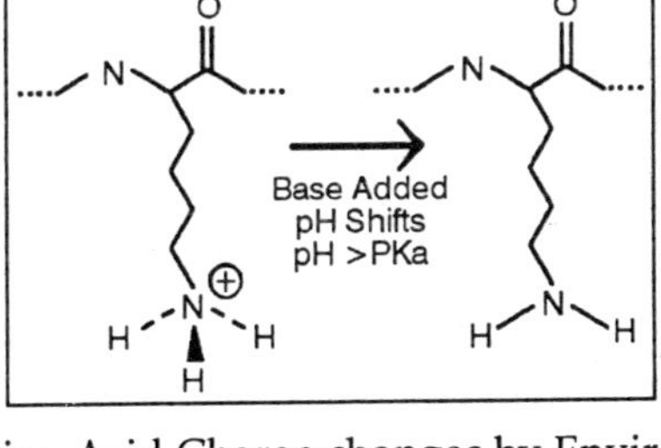

Fig. Amino Acid Charge changes by Environment.

Proteins consist of long chains of amino acids, many of which include acidic or basic groups. These groups are chaged at physiological pH, and form ionic interactions with each other that contribute to the stability of the folded protein.

A change in pH of the protein's aqueous environment may alter the charge of some of these charged groups, disrupting the ionic interactions and the stability of the protein's structure.

Remember that the Henderson-Hasselbalch equation was used for calculating the pH of a solution containing a weak acid and its conjugate base. Since such a solution is a buffer, the Henderson-Hasselbalch equation is extremely useful for calculations involving buffers, allowing the following:

- Determination of the pH of buffered solutions.
- Determination of the ratio of conjugate base to conjugate acid at a given pH.

Example:

A solution of lactic acid has twice as many conjugate acid molecules as conjugate base molecules. If the pKa of lactic acid is 3.85, what is the pH of such a solution?

Answer:

We are given that the concentration of conjugate acid is twice the concentration of conjugate base. Thus,

$$[acid] = 2 [base]$$

Substituting into the Henderson-Hasselbalch equation:

$pH = pK_a + \log ([base]/[acid])$

$pH = 3.85 + \log ([base]/2[base])$

$pH = 3.85 + \log (0.5)$

$pH = 3.85 + (-0.30)$

$pH = 3.55$

Question:

If a buffer solution contains an acid as one component and a base as the other, can those two componenets be HCL and NaOH?

Answer: Your assumption is correct. A buffer is made of a weak acid and its conjugate base or a weak base and its conjugate acid.

So neither HCl nor NaOH will be a component in a buffer; they are both strong. An example of a buffer is acetic acid (a weak acid) and sodium acetate (a conjugate base of acetic acid as a salt). Incidentally, in the example you gave, HCl in water dissociates to hydronium ions and chloride ions. NaOH in water dissociates to hydroxide ions and sodium ions. When the two are mixed, the hydronium ions and the hydroxide ions combine to form water leaving behind the sodium and chloride ions.

If you have the same amount and concentration of HCl as NaOH you will end up with salt water because sodium chloride is simply table salt.

BUFFERS AND IONIZATION OF AMINO ACIDS

- Buffers resist change. pH buffers resist change in pH when acid (H+) or base (OH–) is added.
- pH = –log [H+] –where bracket indicates molar concentration. If [H+] = 10^–7 M, then pH = 7. This is neutral pH since [H+] = [OH–] and it is middle of pH scale. Draw a pH scale by calculating the pH of 1 M HCl and 1 M NaOH. Use the equation Kw = 10^–14 = [H+] × [OH–], to find the pH of 1 M NaOH.
- pH of a buffered solution can be found by using the Henderson-Hasselbach equation:

$$pH = pK + \log([A-]/[HA])$$

All acids and bases have a conjugate acid (HA) and a conjugate base (A–). At [HA] = [A–], pH = pK (prove this with the above equation).

At the pK, a buffer maximally resists change in pH and...the buffering zone is considered to be 1 pH unit above and below the pK.

CONCEPTS

Buffering zone = pK +/–1 pH unit.
At pH = pK, [HA] = [A–].
At pH below pK, [HA] > [A–].
At pH above pK, [HA] <[A–].

- Amino acids (AAs) are more complex than simple buffers and...have at least two pK values because AAs have at least two ionizing groups:

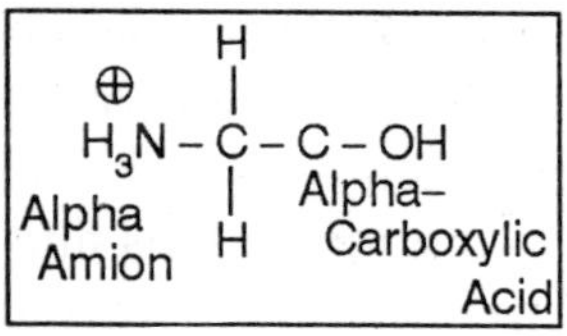

Glycine Structure

Example: Glycine has an alpha-amino group and alpha-carboxylic acid group.

Each ionizable group has a pK (called pK–1 and pK–2).

Glycine pK–1 = 2.3 (alpha-carboxylic acid); pK–2 = 9.6 (alpha amine).

Draw a titration curve for Glycine for practice. In this class, we treat all ionizing groups with in a molecule, as acting independently of other ionizing groups.

The pH on the titration curve where an AA has no net charge is called the pI or isoelectric point. The pI is calculated by averaging the two pK values on either side of the neutral form (ie form with no net charge).

All AAs must have at least 3 ionic forms:"AA+1","AA0", &"AA-1".

Where"AA+1" = AA form with net charge of +1,

"AA0" = neutral form with no net charge, &

"AA-1" = AA form with net charge of –1.

While an AA may have other ionic forms, only these 3 forms count when finding pI:

AA+1 goes to AA0 via pK–below.

AA0 goes to AA-1 via pK–above.

Idea

pI = average of (pK-below) + (pK–above).

where pK-below is the pK between AA+1 and AA0.

and pK-above is the pK between AA0 and AA–1.

5. For more complex AAs, like Asp, Glu, Lys, Arg & His - with 3 ionizing groups, Other forms exist with charges other than +1, zero, & -1.

For these AAs, the best approach is to start with the fully protonated form (ie. One with all the protons it can take on its ionizable groups). Calculate the net charge on this group (it must be positively charged), then titrate the AA to remove the first proton and find net charge again. Continue doing this until all the protons have been removed. Keep track of this by making a simple model. Then, find the form with no net charge (ie. AA0) and use the pK values, which govern the transition from AA+1 to AA0 and AA0 to AA-1, as the ones to calculate the pI.

His has 3 ionizing groups, alpha-carboxylic acid (pK 1.8), side-chain amino (pK 6.0) and alpha-amino (pK 9.2):

His+2 goes to His+1 via pK 1.8;
His+1 goes to His0 via pK 6.0;
His0 goes to His-1 via pK 9.2.
Therefore, pI = (6.0 + 9.2)/2 = 7.6.

- Some of you have trouble figuring out which group ionizes first. Here are some rules to apply to help you.
  - Carboxylic acids ionize at acidic pH; ie. Carboxylic acids give up their protons at acid pHs.
  - Amino groups ionize at basic pH; ie. Amines give up their protons at basic or alkaline pHs.
  - When groups with a similar chemical nature are present:
    a. Carboxylic acids near an amino group have a more acidic pK than isolated carboxylic acids.
    b. Amino groups near a carboxylic acid have a more acidic pK than isolated amines.
    c. Aromatic amines (like in His side chain) have a pK near neutral.

Apply the above rules and concepts to calculate the pI for all the other complex AAs. Asp, Glu, Lys, and Arg. Their pK values and structures are shown below:

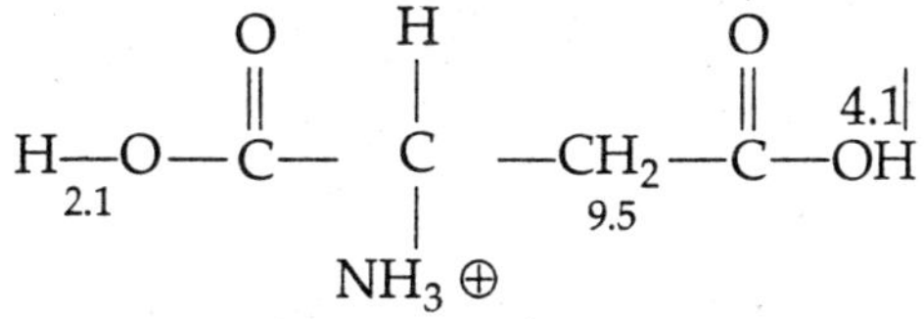

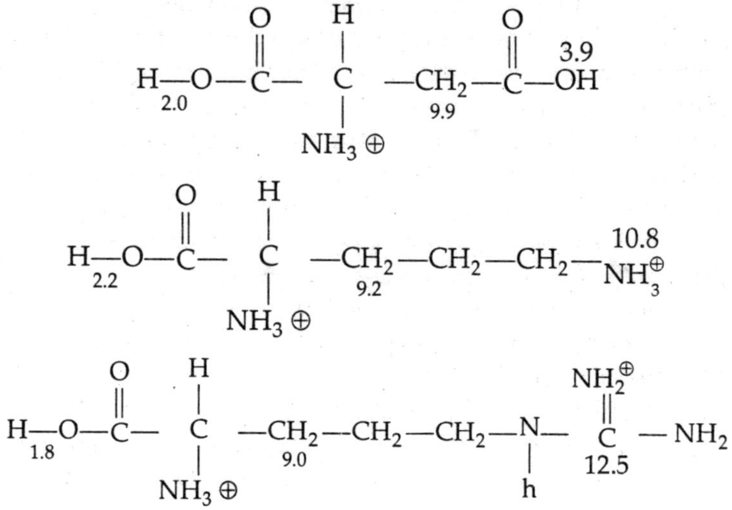

BIOLOGICAL BUFFERS

Many biochemical processes are markedly impaired by even small changes in the concentrations of free H+ ions. It is therefore usually necessary to stabilise the H+ concentration in vitro by adding a suitable buffer to the medium, without, however, affecting the functioning of the system under investigation. A buffer keeps the pH of a solution constant by taking up protons that are released during reactions, or by releasing protons when they are consumed by reactions. The observation that partially neutralised solutions of weak acids or bases are resistant to changes in pH when small amounts of strong acids or bases are added led to the concept of the'buffer'.

The Biological Buffers imparts basic knowledge on biological buffer substances and support in terms of selecting the best buffer substance for your assay system. Be aware of the fact that some buffers serve as substrate/ inhibitor of enzymes (*e.g.* phosphate) or the temperature sensitivity of the pH of some buffers, especially Tris, probably the most frequently employed buffer substance.

BUFFER CONCEPT

Many biochemical processes are markedly impaired by

even small ch anges in the concentrations of free H+ ions. It is Therefore usually necessary to stabilise the H+ concentration in vitro by adding a suitable buffer to the medium, without, however, affecting the functioning of the system under investigation.

A buffer keeps the pH of a solution constant by taking up protons that are released during reactions, or by releasing protons when they are consumed by reactions.

The observation that partially neutralised solutions of weak acids or bases are resistant to changes in pH when small amounts of strong acids or bases are added led to the concept of the'buffer'.

BUFFER CAPACITY

Buffers consist of an acid and its conjugated base. The quality of a buffer is determined by its buffer capacity, *i.e.* its resistance to changes in pH when strong acids or bases are added. In other words: the buffer capacity corresponds to the amount of H+ or OH- ions that can be neutralised by the buffer. The buffer capacity is related to the buffer concentration. The graph described by the relation of the pH to the addition of H+/OH– ions is called the titration curve.

The point of inflection of the curve corresponds to the pKa value. At this point, the buffer capacity is at its maximum at the pKa value. This point therefore corresponds to the mid-point of the pH range covered by the buffer and is where the concentration of acid and base is the same.

In the area of this pH range, therefore, relatively large amounts of H+/OH– ions result in only small changes in pH. A basic principle is that a buffer that has a pH value of one pH unit above or below the pKa value loses so much buffer capacity that it no longer has any real buffer function.

Based on the Henderson-Hasselbalch equation pH = pKa + log [A–]/[HA] for the calculation of the pH of a weak acid or alkaline solution, the ions in the water must also be taken into account when working in pH ranges below 3.0 and above 11.0. Most biochemical reactions, however, take place in the pH range between 6.0 and 10.0.

THE PH VALUE

Conductivity can also be detected in highly purified water due to the OH– and H3O+ ions resulting from the autoprotolysis of water. This intrinsic dissociation of the water is an equilibrium reaction and the product of the concentrations of the two ions represents a constant:

$$K = [H3O+] \times [OH-].$$

This value is depends on the temperature only and is 10-14 for purified water at 22°C. Depending on which of the two ions is present in a higher concentration in a solution, the solution is called to be acidic or alkaline. To express this fact in terms of a simple number, the negative exponent of the easily measurable hydronium ion concentration [H3O+] was chosen.

This dimensionless number is called the pH value. The pH can also be described as the negative, decadic logarithm of the hydronium ion concentration of a solution:

$$pH = -\log [H3O+]$$

The hydronium ion concentration of pure water is 10–7 mol/L, as can be derived from the above equation for the ion product. Therefore, the pH value is 7. In an acidic solution, the concentration of H3O+ ions is increased (*e.g.* from 10–7 mol/L to 10–2 mol/L) and the pH is therefore < 7. In an alkaline solution, the hydronium ion concentration is decreased and this results in a pH > 7.

THE PKA VALUE

To change the pH value of a solution, substances are dissolved in the solution which release H+ ions into the water (acids), thus raising the H3O+ ion concentration and lowering the pH value, or which decrease the H3O+ ion concentration and thus increase the pH value by taking up H+ ions (bases).

As always in chemistry, this reaction is an equilibrium reaction, and the capacity of a substance to shift this equilibrium in one direction or another is determined by the potency of the acid. It is calculated from the equilibrium constants using the following equation $K = [A] \times [H3O+]/ [HA]$, requirements.

DILUTION AND RINSE BUFFERS

Rowmanowsky stains are pH sensitive. Tinctorial (stain picture) quality can be altered by changes in the final pH of staining reagent, this can be caused by differences in the quality or source of the water used to dilute such stains. Tinctorial quality and consistency can be maintained by the use of dilution buffers, which when added to the diluent water will produce stain reagents of the required pH.

Acid pH (<7.0) will emphasize eosinophilic staining.

Basic pH (>7.0) will emphasize azureophilic (basophilic) staining.

In general, blood stains are diluted to produce solutions of slightly acid pH, 6.50-6.80.

We offer a range of buffer concentrates covering pH's 6.50/ 6.80/ 7.00/ 7.20/ 7.40/ 7.60, supplied in 10 vial or 25 vial packs each vial makes 5 litres of dilution buffer. They offer great value for money and the certainty of preparing blood stains of the required pH time and again. To maintain the tinctorial qualities of the final stain slides should be rinsed in the same buffer. Concentrates - ready-to-use. High quality, accurate, time saving and good value for money.

WATER

Water has many unique properties that make it essential to all life. Most of water's unique properties are a result of the hydrogen bonding between water molecules.

Water is an excellent solvent. When ionic compounds are placed into water, the ions dissociate or separate. Polar covalent compounds, because they too have charged poles, also dissolve in water. Nonpolar covalentcompounds, however, do not dissolve in water. Thus polar covalent compounds are hydrophilic (water loving) while nonpolar covalent compounds are hydrophobic (water fearing).

Water has high cohesion. Individual water molecules tend to"stick" with other water molecules due to hydrogen bonding. This leads to two characteristics of water: a high surface tension and strong capillary action. Water's high surface tension is what allows some insects to run across the surface of the water.

It is also the reason why belly-flops hurt so much. Water's strong capillary action is what allows the liquid level in a straw to be higher than that in the surrounding drink. Using hydrogen bonding, water molecules attract others up the sides of the straw. This effect is the more noticeable in thinner straws.

You're right if you think it's terribly difficult to select an aspect or two to discuss when the larger topic is water. There's all sorts of fascinating stuff! Everything from the role of the waters of hydration in antibody-binding to the effect of solvation for stabilizing protein structure to the dependence of water's dielectric and pH on temperature and pressure. This last topic is particularly exciting as we learn more about the thermophiles living at deep sea vents. And to paraphrase a guest I heard on a radio science-talk show recently, once we better understand the life living in our oceans, we'll be much better prepared to understand possible life on other planets.

But most importantly, because the aqueous environment influences all of the chemistry that takes place in it, when you are discussing water you get to address just about anything, while still ostensibly focusing on the substance itself.

Before we get to water at extremes of temperature and pressure, let's touch briefly on the oh so neglected role of water in our everyday life. And just to be sure we start out on a provocative note, let me here quote Felix Franks who told me

BIOCHEMISTRY AS CHEMISTRY OF WATER

We immediately see the evident truth in this statement when we consider the complex thermodynamics of protein and peptide structural stability and the prominent role of solvent in influencing this equilibrium. Proteins unfold from their native fold upon the addition of urea or other denaturants, peptide helicity increases upon addition of TFE, and alcohol is even known to induce helical structure in regions that otherwise adopt beta-sheet conformations. While the helix inducing effects of alcohol upon peptide conformation have long been used to probe and enhance latent structure, thereby making it measurable by CD or NMR, a recent paper illustrates how this dependence of structure on solvent conditions may be harnessed to make a sort of'conformational switch'. Successful molecular design or peptide engineering applications will necessarily include an assessment of solvation effects for conformational stability.

Turning from a macroscopic picture of the role of water and aqueous solutions in establishing the equilibria of proteinaceous substances to a more microscopic view, we can consider the surprising role played by discrete water molecules in antigen-antibody recognition. In a 1993 paper in Current Opinion in Immunology, Mariuzza and Poljak report a case in which interfacial water plays a critical role in determining the complementarity of the antigen and antibody surfaces.

This work was illuminating, since previous work was leading to the conclusion that the packing density was lower at protein interfaces than in protein interiors. It also appeared that polar atoms at these interfaces were apparently tolerating the absence of neutralizing partners - far more than they do in protein interiors. In this work, a 1.8 angstrom crystal structure of an immune complex was studied and revealed that the water network was an integral part of the interface.

As pointed out by these authors, the analysis of low resolution crystal structures may in fact lead to erroneous conclusions on the nature of the interface in these complexes. Furthermore, the evidence reported in this work points to the inescapable conclusion that the intermolecular complementarity essential for molecular recognition is not necessarily imparted solely by the large molecules.

And, contrary to what our notions of solvent entropy might lead us to expect, bound water may play a role in stabilizing the structure of single proteins as well as that of complexes. And, this bound water may, apparently, serve this function at the surface of the protein, and not solely while confined to the interior. That's just the tip of the iceberg as far as far as water and its impact for biological systems at what is considered physiological temperature and pressure; the realm normally explored by biochemists and pharmacologists.

Much, however is being learned about the so-called"hot life", life found at thermal vents deep in the ocean floor.

Reports have documented the existence of organisms at temperatures in excess of 110 deg. C. These organisms are referred to collectively as thermophiles. It is important to note the enormous temperature variation of sea water at these depths, with a documented range from 2 deg.

C to 900 deg. C at depths of a few km where the high pressure precludes boiling. Hot life, or the existence of thermophiles, has been noted at temperature in excess of 100 deg. C at these elevated pressures, and it remains a matter of some dispute whether or not bacteria can exist at even higher temperatures still.

So, what are the properies of ambient sea water for organisms at deep sea thermal vents? We can expect these properties to influence the kinetics and thermodynamics of the biochemistry pertinent to this life, just as they influence the systems with which we are, perhaps, more familiar. We can begin to understand the fluid environment presented to thermophiles by considering the temperature dependence of a few properties of pure water.

A quick look at the following data from Shock show the dramatic effect of temperature on water's dielectric. Water at 25 deg. C and 1 atm pressure has the familiar dielectric near 80; however near water's critical point the dielectric has fallen to less than 10. For comparison, ethanol has a dielectric of roughly 25 at STP.

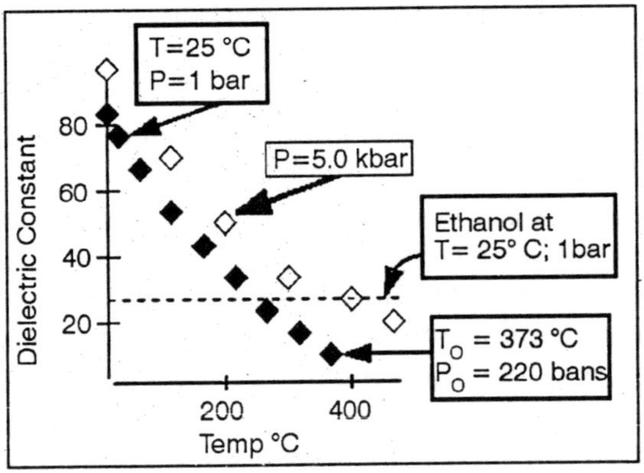

Fig. From Shock.

In this graph all white and blue diamonds pertain to water data points. The blue diamonds are for pressures below the critical point where water exists in two phases, while the white diamonds pertain to 5 kbar data. For the low pressure data, two key data points are indicated. The grey dashed line indicates the dielectric of ethanol at standard conditions. (This graph is drawn to approximate that shown in the original paper which should be consulted for quantitative work.)

The weak pressure dependence of this data should be noted. In order to understand the relevance of this sort of analysis to a complete understanding of ambient conditions at deep sea thermal vents, it should be understood that the pressure at these depths ranges from a few hundred bars to several kilobars. Shock should be consulted for additional details, data and a general groovy time.

Well, if all this is true, then water must have a pretty different structure at these T&P's than we are used to. Right? Not surprisingly, this is what is being found. Some of these results are shown in this graph:

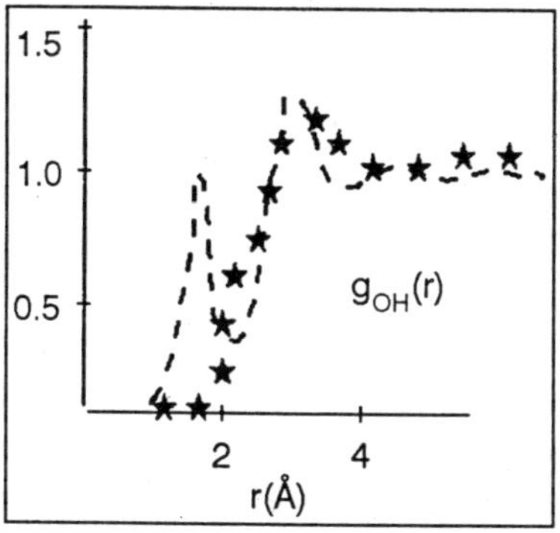

Fig. From Postorino

The pair distribution function for intermolecular OH atom pairs is shown as a function of interatomic distance for water at 25 deg. C (dashed line) and 400 deg. C.

This particular pair distribution function tells us the likelihood with which a H atom will be found at any distance from an O atom (The intramolecular data isn't shown here, but the water molecule does have the same structure at 400 deg. C as at 25 deg. C. This graph considers only intermolecular associations.) Immediately, then, we see that at room temperature it is relatively likely to find these atom pairs separated by 1.95 angstroms. This is the hydrogen bond. And then there are strong secondary structural phenomena, manifest as the damped oscillations that we see here out to beyond 5 angstroms, all due to the overwhelming hydrogen bonding Behaviour.

Looking at the high temperature data, it's clear that the hydrogen bonding peak, which dominates the room temperature data, is completely gone. The OO correlation function, not reproduced here, shows a slight broadening of the nearest-Neighbour peak with temperature. The high temperature OO pair correlation function also yields a larger coordination number than is seen at room temperature. In other words, each water molecule is surrounded by more water molecules in its first solvation shell at 400 deg.

C than at 25 deg. C. At 400 deg. C the packing is not restricted by the severe hydrogen-bonding geometrical constraints of the the room temperature fluid. This result is in line with the close packing Behaviour of most liquids. These results taken together with those for the dielectric phenomena described above begin to paint a picture of high temperature water. Water at high temperature exhibits many of the properties we associate with less polar liquids under standard conditions.

In order to understand something of the biochemistry that might pertain under extreme aqueous conditions, Shock has been estimating the stability of peptidic species at elevated temperatures and pressures. Specifically he has been asking questions like"How does the free energy of the peptide bond change with temperature and pressure?"

A snapshot view of some of this work will serve to further the discussion of aqueous properties under extreme conditions while at the same time giving a hint for some of the very different biochemistry that will undoubtedly be revealed by further study in this area.

Considering only Glycine here, the free energy of interest here is that which determines the following equilibrium:

Glycine + Glycine = Glycylglycine + H_2O

In other words, what is the free energy cost to form the peptide bond between two Glycine molecules? This equilibrium can be calculated from the free energies of formation for the individual species involved in this reaction, Glycine and Glycylglycine, and the temperature dependence of this free energy in the following way:

$$\Delta G^{\circ}_{PT} = \Delta G^{\circ}_{f} - S_{Pr,Tr}(T - Tr) + \int C^{\circ}_{P}$$

$$-T\int C^{\circ}_{P} d\ln T + \int V^{\circ} dP$$

This equation has the usual host of characters. G represents free energy, S is entropy, Cp is heat capacity, V is volume and P is pressure. If any of this is murky please consult the original paper by Shock.

What we're basically doing is getting the free energy (OK, so it's really the standard partial molal Gibbs free energy) at any T&P from that at STP by integrating over the temperature and pressure dependence of the heat capacity and volume.

In order to do this, we need to know something about the way heat capacity and volume depend on the thermodynamic variables of T and P, respectively.

We'll look at this first:

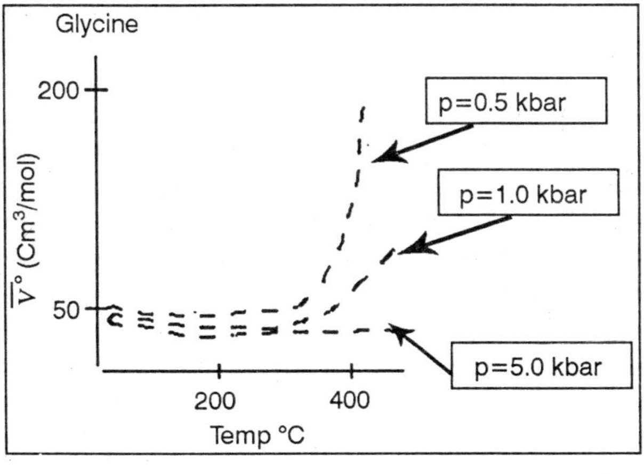

Standard partial molal volume for Glycine. From Shock. (This graph is drawn to approximate that shown in the original paper which should be consulted for quantitative work.)

These results show the very large dependence of the molal volume on temperature at lower pressures, and the much weaker temperature dependence at extremely high pressures.

Here are the results for the heat capacity:

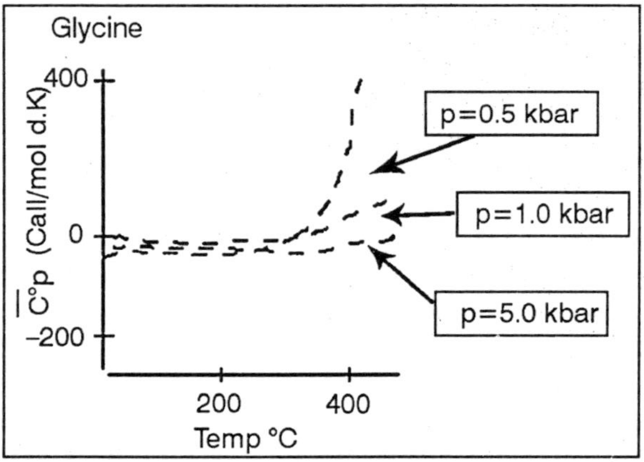

Standard partial molal heat capacity (constant pressure variety) for Glycine. From Shock. (This graph is drawn to approximate that shown in the original paper which should be consulted for quantitative work.)

Pretty much the same thing. A large T dependence is witnessed at lower temperature, but this Behaviour is convincingly squashed at high pressure.

The results for Glycine have been shown here. Analogous results for Glycylglycine are obtained from these by calculating a regression equation that relates, for instance, the partial molal volume of Glycine to that of Glycylglycine.

Putting it all together with some other necessary data (like the free energy of formation at STP, and the entropy at STP; you got it, Shock).

The free energies of formation for both Glycine and Glycylglycine can be obtained at any temperature and pressure, and from these the free energy governing the equilibria may be directly obtained.

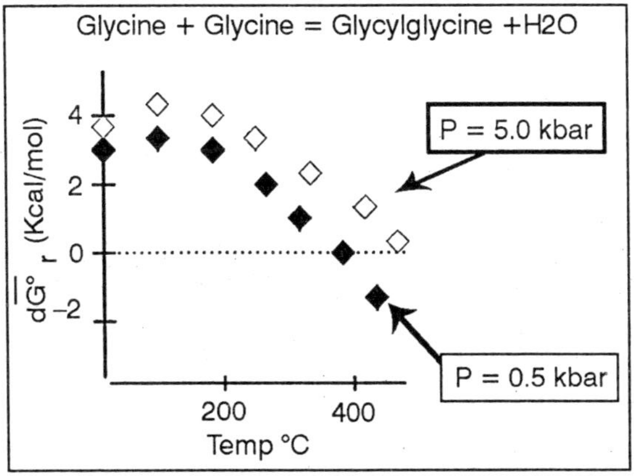

Free energy required to form the Glycylglycine dipeptide from the constituent Glycine moieties. From Shock. (This graph is drawn to approximate that shown in the original paper which should be consulted for quantitative work.)

The results indicate that while it costs free energy to form the dipeptide at lower temperatures and pressures (at STP the free energy of this reaction is roughly 3.5 kcal/mol), at extremes of temperature and pressure the dipeptide is the thermodynamically favoured species. In any event, these results are consistent with those for many other dipeptides not covered here and predict that peptide bonds become more stable at elevated temperatures. This Behaviour just begins to hint at the interesting biochemical phenomena that will be revealed for life in water under extreme conditions where water has very different properties from those to which we are accustomed.

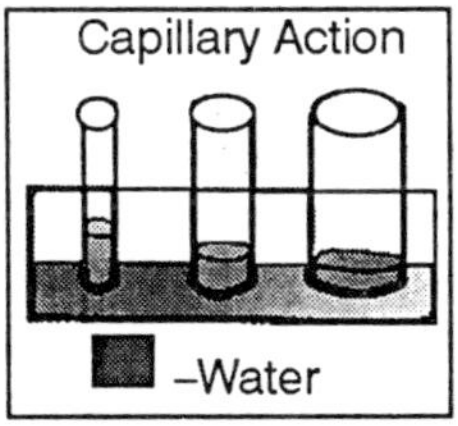

Water has a high specific heat capacity. It takes a lot of energy to change the temperature of water. This characteristic makes water a good insulator and a good coolant. When you sweat, your body is using water as a coolant. The eVapourating water removes heat with it. At night, oceans are a good insulator. The energy that the sun spent in heating the water all day is slowly released into the night.

Finally, water is the most dense at 4 degrees Celsius. Hydrogen bonding arranges water molecules into hollow"cells" when water freezes, making it less dense than liquid water. This characteristic of water is what allows fish in lakes and ponds to survive in the winter. When the water freezes, it becomes less dense and floats to the surface -- leaving the bottom of the lake or pond unfrozen.

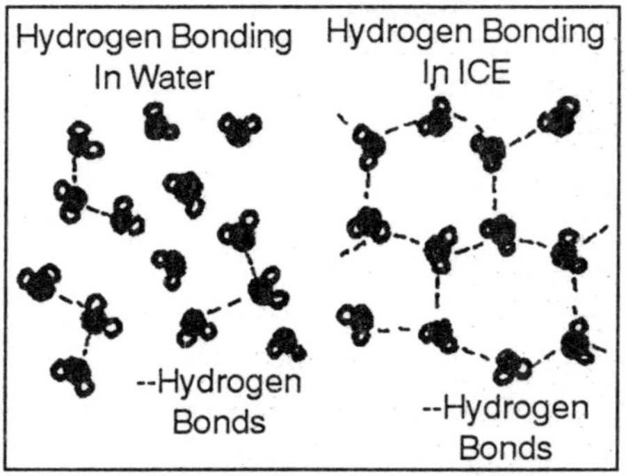

WEAK INTERACTIONS IN AQUEOUS SYSTEMS

Water is a good solvent for polar solutes with which it forms hydrogen bonds. Non-polar compounds dissolve poorly in water. They can not form a hydrogen bond with the solvent. Numerous weak non covalent interactions influence the folding of macromolecules.

The physical properties of aqueous solutions are strongly influenced by the concentration of solutes. Ex: The tendency for water to move across a semi permeable membrane is the osmotic pressure.

IONIZATION OF WATER

Pure water is slightly ionized. pH is defined as the negative logarithm of the molar hydronium ion concentration. An acid that ionizes completely in water is a strong acid. An acid which is partially ionized in water is a weak acid. pk provides the measure of the ease of the dissociation of the proton. The lower the number the stronger is the acid.

BUFFERING AGAINST PH

Buffers are aqueous systems that tend to resist in changes in pH when small amounts or acids or base are added. It consists of a weak acid and its conjugate base. As an example a mixture of equal concentrations of acetic acid and acetate ion is a buffer system. Every life form is extremely sensitive to slight pH changes. Human blood for example needs to remain within the pH range of 7.38 to 7.42. The relationship between pH and pk is described by Henderson - Hasselbach equation. It is used to calculate the pH of biological fluids. Two especially important biological buffers are the phosphate and bicarbonate systems. The property of a solution that depends on the concentration of a dissolved solute but not on its chemical identity is known as colligative properties. Substances that do not ionize in water are non electrolytes. Vapour pressure, boiling point, freezing point and osmosis are colligative properties.

WATER AS A REACTANT

Water is very often a direct participant in the chemical reactions of living cells. Formation of ATP from ADP and inorganic phosphate is an example of a condensation reaction in which the elements of water are eliminated. Two amino acid molecules are covalently joined through a peptide bond with a loss of water molecule.

FITNESS OF AQUEOUS ENVIRONMENT FOR LIVING ORGANISMS

The high specific heat of water is useful to cells and organisms because it allows water to act as a'heat buffer', keeping the temperature of an organism relatively constant as the temperature of the surroundings fluctuates.

SUMMARY

Water is a good solvent for polar solutes with which it

forms hydrogen bonds. Non-polar compounds dissolve poorly in water. They can not form a hydrogen bond with the solvent. Numerous weak non covalent interactions influence the folding of macromolecules. The physical properties of aqueous solutions are strongly influenced by the concentration of solutes.

The equilibrium constant describes whether an acid or a base is strong or weak. A mixture of weak acid and its salt resists changes in pH caused by the addition of H+ or OH-. The mixture functions as a buffer. The relationship between pH and pk is described by Henderson-Hasselbach equation. It is used to calculate the pH of biological fluids. Water is very often a direct participant in the chemical reactions of living cells

PROPERTIES AND BIOMOLECULAR STRUCTURE

Water is necessary for life. Many plant and animal adaptations conserve water - the thick skin of desert cacti and the intricate structure of the mammalian kidney are just two examples. Planetary scientists look for evidence of liquid water when speculating about the possibility of life on other planets such as Mars or Jupiter's moon, Titan.

WATER REMARKABLE PROPERTIES

High Surface Tension

Despite being denser than water, small objects, such as aquatic insects, can stay on top of water surface.

High Boiling Point

Relative to its molecular weight, water boils at a high temperature. For example, ammonia, with a molecular weight of almost 17, boils at –33° C, while water, with a molecular weight of 18, boils at 100° C.

Density is Dependent on Temperature

Solid water (ice) is less dense than liquid water. This

property means that lakes and ponds freeze from the top down, a benefit to the fish living there, who can overwinter without being frozen solid.

PROPERTIES OF WATER AND HYDROGEN BONDS

Water has a dipole, that is, a separation of partial electrical charge along the molecule. Two of oxygen's six outer-shell electrons form covalent bonds with the hydrogen. The other four electrons are nonbonding and form two pairs. These pairs are a focus of the partial negative charge, and the hydrogen atoms correspondingly become partially positively charged. Positive and negative charges attract each other, so that the oxygen and hydrogen atoms form hydrogen bonds.

Each oxygen in a single molecule can form H-bonds with two hydrogens (because the oxygen atom has two pairs of nonbonding electrons). Figure shows such a hydrogen bond. The resulting clusters of molecules give water its cohesiveness. In its liquid phase, the network of molecules is irregular, with distorted H-bonds. When water freezes, the H-bonds form the water molecules into a regular lattice with more room between the molecules than in liquid water; hence, ice is less dense than liquid water.

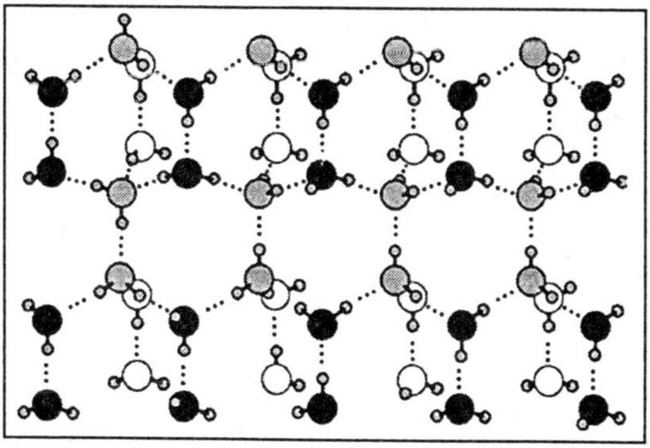

HYDROGEN BONDS AND BIOMOLECULES

In water, the nonbonding electrons are the H-bond acceptors and the hydrogen atoms are the H-bond donors. Biomolecules have H-bond acceptors and donors within them.

Consider the side chain of a simple amino acid, serine. The oxygen contains two pairs of nonbonding electrons, as water does, and the hydrogen is correspondingly a focus of partial positive charge. Serine thus can be both an H-bond acceptor and donor, sometimes at the same time.

As you would expect, serine is soluble in water by virtue of its ability to form H-bonds with the solvent around it. Serine on the inside of a protein, away from water, can form H-bonds with other amino acids; for example, it can serve as an H-bond donor to the nonbonding electrons on the ring nitrogen of histidine, as shown in Figure.

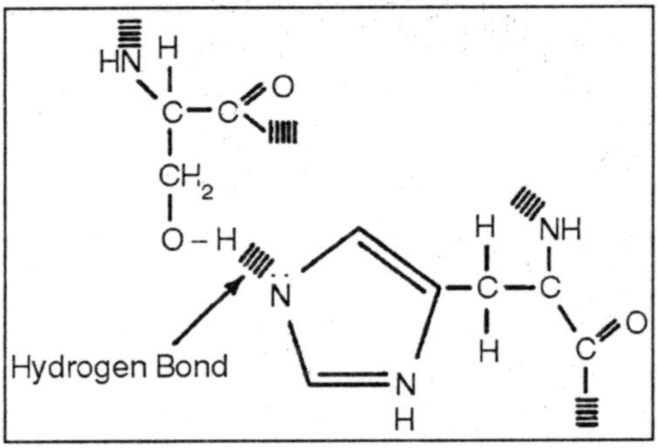

These H-bonds normally exist only when water is not present. If serine's side chain is found on the surface of a protein, it is very likely to form H-bonds, given the relatively high concentration of water available.

THE POLARITY OF WATER

Water has a simple molecular structure. It is composed of one oxygen atom and two hydrogen atoms. Each hydrogen atom is covalently bonded to the oxygen via a shared pair of electrons. Oxygen also has two unshared pairs of electrons. Thus there are 4 pairs of electrons surrounding the oxygen atom, two pairs involved in covalent bonds with hydrogen, and two unshared pairs on the opposite side of the oxygen atom. Oxygen is an"electronegative" or electron"loving" atom compared with hydrogen. Water is a"polar" molecule, meaning that there is an uneven distribution of electron density. Water has a partial negative charge (δ–) near the oxygen atom due the unshared pairs of electrons, and partial positive charges (δ+) near the hydrogen atoms. An electrostatic attraction between the partial positive charge near the hydrogen atoms and the partial negative charge near the oxygen results in the formation of a hydrogen bond as shown in the illustration.

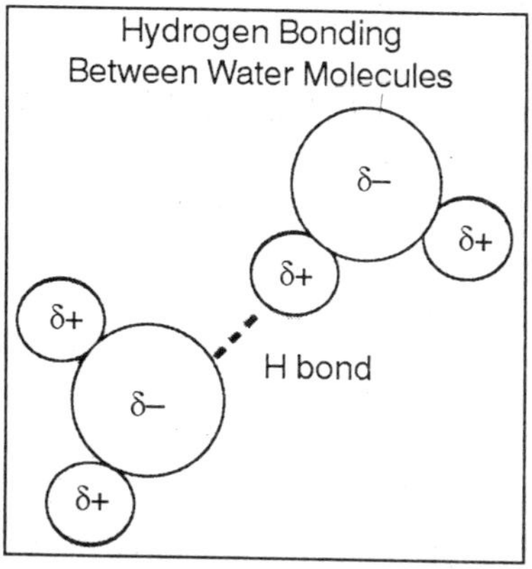

The ability of ions and other molecules to dissolve in water is due to polarity. For example, in the illustration below sodium chloride is shown in its crystalline form and dissolved in water.

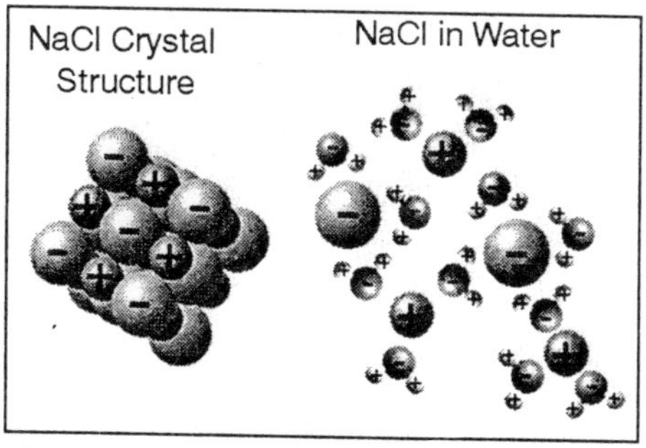

Many other unique properties of water are due to the hydrogen bonds. For example, ice floats because hydrogen bonds hold water molecules further apart in a solid than in a liquid, where there is one less hydrogen bond per molecule. The unique physical properties, including a high heat of Vapourization, strong surface tension, high specific heat, and nearly universal solvent properties of water are also due to hydrogen bonding. The hydrophobic effect, or the exclusion of compounds containing carbon and hydrogen (nonpolar compounds) is another unique property of water caused by the hydrogen bonds. The hydrophobic effect is particularly important in the formation of cell membranes. The best description is to say that water"squeezes" nonpolar molecules together.

ACIDS AND BASES, IONIZATION OF WATER

$$H_2O \xrightarrow{\text{Limited}} H^+ + OH^-$$

- Acid release H^+
- Bases accept H^+

We define the pH of a solution as the negative logarithm of the hydrogen ion concentration:

- At pH 7.0, a solution is neutral.
- At lower pH (1-6), a solution is acidic.
- At higher pH (8-14), a solution is basic.

3

Drugs

INTRODUCTION

Toxic side effects of local anaesthetic drugs occur when excessive blood levels occur. This is usually due to.

- Accidental rapid intravenous injection.
- Rapid absorption, such as from a very vascular site ie mucous membranes. Intercostal nerve blocks will give a higher blood level than subcutaneous infiltration, whereas plexus blocks are associated with the slowest rates of absorption and therefore give the lowest blood levels.
- Absolute overdose if the dose used is excessive.

REDUCING THE RISK OF TOXICITY

- Decide on the concentration of the local anaesthetic that is required for the block to be performed. Calculate the total volume of drug that is allowed according to the table below.
- Use the least toxic drug available.
- Use lower doses in frail patients or at the extremes of ages.
- Always inject the drug slowly (slower than 10ml/minute) and aspirate regularly looking for blood to indicate an accidental intravenous injection.
- Injection of a test dose of 2-3ml of local anaesthetic containing adrenaline will often (but not always) cause a significant tachycardia if accidental intravenous injection occurs.

- Most nerve blocks are more dependent on volume of drug injected than the total dose. Therefore if more volume is needed it is better dilute the local anaesthetic with 0.9% saline than to add more local anaesthetic and increase the dose unnecessarily.
- Add adrenaline (epinephrine) to reduce the speed of absorption. The addition of adrenaline will reduce the maximum blood concentration by about 50%. Usually adrenaline is added in a concentration of 1:200,000, with a maximum dose of 200 micrograms.
- Make sure that the anaesthetist or a trained nurse monitors the patient closely during the administration of the local anaesthetic and the following surgery.

Drug	**Maximum dose for Infiltration**	** Maximum dose for Plexus Anaesthesia**
Lignocaine	4mg/kg	5mg/kg
Lignocaine with adrenaline	7mg/kg	7mg/kg
Bupivacaine	2mg/kg	2mg/kg
Bupivacaine with adrenaline	3mg/kg	3mg/kg
Prilocaine	6mg/kg	7mg/kg
Prilocaine with adrenaline / octapressin	8mg/kg	8mg/kg
* when performing intercostal blocks, reduce the dose for infiltration by 25%		

SIGNS AND SYMPTOMS OF LOCAL ANAESTHETIC TOXICITY

The systemic toxic effects due to local anaesthetic overdose primarily involve the central nervous and cardiovascular systems. In general the Central Nervous System (CNS) is more sensitive to local anaesthetics than the Cardiovascular System (CVS). Therefore CNS manifestations tend to occur earlier. Brain excitatory effects occur before the depressant effects.

CVS SIGNS & SYMPTOMS

- *Early or mild toxicity*: Tachycardia and rise in blood pressure. This will usually only occur if there is adrenaline in the local anaesthetic. If no adrenaline is added then bradycardia with hypotension will occur.
- *Severe toxicity:* Usually about 4 – 7 times the convulsant dose needs to be injected before cardiovascular collapse occurs. Collapse is due to the depressant effect of the local anaesthetic acting directly on the myocardium. Bupivacaine is considered to be more cardiotoxic than lignocaine. Severe and intractable arrhythmias can occur with accidental iv injection.

The acute toxicity of local anaesthetics is due to the speed of rise of blood concentration. Therefore a rapid injection of a small volume may cause toxicity.

ESSENTIAL PRECAUTIONS

- Secure intravenous access before injection of any dose that may cause toxic effects.
- Always have adequate resuscitation equipment and drugs available before starting to inject.

TREATMENT OF TOXICITY

If a patient you are attending shows any signs or symptoms of toxicity during injection of local anaesthetic stop the injection and assess the patient.

Treatment is based on the A B C D of Basic Life Support Call for help while treating the patient:

A. Ensure an adequate airway; give oxygen in high concentration if available.
B. Ensure that the patient is breathing adequately.

Ventilate the patient with a self-inflating bag if there is inadequate spontaneous respiration. Intubation may be required if the patient is unconscious and unable to maintain an airway.

C. Treat circulatory failure with intravenous fluids and vasopressors such as ephedrine (10mg boluses) if hypotension occurs. Adrenaline may be used cautiously intravenously in boluses of 0.5 - 1ml of 1:10,000 (1mg in 10ml) if ephedrine is either not available or not effective in correcting the hypotension. Treat arrhythmias.Start chest compressions if cardiac arrest occurs.

D. Drugs to stop fitting such as Diazepam 0.2-0.4mg/kg intravenously slowly over 5 minutes repeated after 10 minutes if required, or 2.5mg - 10 mg rectally. Thiopentone 1-4 mg/kg intravenously may also be used in theatre.Observe the patient closely after any reaction.

Treatment of local anaesthetic toxicity is likely to have a good outcome if toxicity is recognised and basic resuscitation is started early. Monitor patients closely when using local anaesthetics.

If a reaction occurs:

- Prevent hypoxia, which will cause brain damage and make fitting or arrhythmias more difficult to control.
- Ensure that hypotension and arrhythmias are treated early.
- Ensure that fits are adequately treated.
- Most reactions are short-lived if the above advice is followed.

Although there are a wide range of recognised poisons and toxic substances, it is important to remember that some of the most well-known poisonous substances are in fact drugs that are presently used worldwide to treat various symptoms and cure diseases. Due to the complex nature of the human body, it is practically guaranteed that all useful drugs will produce unwanted effects, but some can produce positively dangerous effects. Toxicity refers to these sometimes-deadly effects: though all drugs have adverse effects at normal doses, it is only the dangerous ones, which are referred to as toxic.

Drugs may have toxic effects on various organs, with some potentially useful drugs being abandoned due to toxic effects, and overdoses of some existing drugs leading to death. Doctors are often placed in difficult positions while trying to decide whether the benefits of a drug outweigh its risks and must be forever vigilant in looking for unknown effects.

A HISTORY OF DRUG TOXICITY

One of the first major incidences of drug toxicity in the 20th Century was in the USA in 1937 and involved a drug known as sulphanilamide, an anti-bacterial agent used to treat streptococcal infections. The drug had been used safely for quite some time in its tablet and powder forms, but it emerged that there was a call for the drug to be available in liquid form. This led to the production of an 'elixir of sulphanilamide', which the pharmacists in the USA found would dissolve in diethylene glycol. With the promise of increased profits, the elixir was immediately mass-produced without testing and sent all over the country. In the space of one month, over a hundred patients died, but it was not the drug that killed them. The solvent, diethylene glycol, is rather similar to common antifreeze (ethylene glycol) and is a deadly poison, but this fact had somehow been ignored.

Thalidomide

Although this sort of occurrence is now avoided by careful testing of new formulae, other drugs can have toxic effects that are much more difficult to detect. One of the most famous examples of a drug prescribed to the public and then hastily withdrawn is thalidomide. Introduced in 1957 in West Germany to treat morning sickness in pregnant women, thalidomide had not been tested thoroughly enough to discover its teratogenic effects (see below). It was only when the pregnant women who had been treated with the drug started giving birth to deformed children with shortened limbs and no external ears that thalidomide was withdrawn.

By then, the drug was being used throughout Europe and had damaged 15,000 foetuses: 12,000 were deformed and half went on to die within the first year of life. Although the use of thalidomide in pregnant women is now banned, a new use for the drug has been found in patients with multiple myeloma and leprosy, and also has applications in Behcet's syndrome, AIDS and tuberculosis. This is partly due to the fact that the drug has been refined so that it only includes the (+) isomer, which has beneficial effects, and not the (-) isomer, which caused the many deformities associated with the drug.

Many More Examples

Although pharmacologists now had the experience provided by the thalidomide tragedy, it still proved difficult to detect the harmful side effects of many drugs. In the four decades that followed, many discoveries were made, all of them after the drug in question was released:

- The contraceptive pill, which in the 1960s contained 75mg of oestrogen, was found to increase the rate of thrombosis to a much higher rate than that found in pregnant women. Using only a third of the quantity of oestrogen has since reduced the risk.
- Clioquinol, an anti-infective agent, was found to cause SMON (subacute myelo-opticoneuropathy), a condition affecting the spinal cord and optic nerve, leading to disturbances of vision and movement as well as mental disorders.
- It was discovered that phenytoin, an anti-epileptic drug, could cause decalcification of bones, leading to rickets in children and osteomalacia in adults. This is due to the drug increasing the production of a hormone that decalcifies bone.
- During the 1970s, the beta-blocker practolol was found to cause an immune reaction known as oculomucocutaneous syndrome, causing inflammation of the eyes, mucous membrane and skin.
- In 1982, benoxaprofen, an NSAID (non-steroidal anti-

inflammatory drug) used to treat arthritis, was released onto the market after tests on rhesus monkeys showed that it was safe. Sixty deaths and 3,500 serious illnesses occurred in the UK alone before the drug was identified as damaging the liver and also causing inflammation when patients were exposed to sunlight.

- In 1984, it was discovered that fenclofenac, another NSAID used to treat arthritis and inflammation, also caused severe liver damage. It had been released to the public after passing tests on ten different species of animal.
- Ten years later, it was found that the NSAID phenylbutazone caused liver and bone marrow damage as well as inflammation on exposure to sunlight. The effects on bone marrow were only discovered in lab cultures of human cells after many cases of aplastic anaemia, a disease where the bone marrow fails to produce sufficient blood cells.
- During the late 1990s, both pemoline, a treatment for ADHD (attention deficit hyperactivity disorder), and troglitazone, a drug used to regulate type-2 diabetes, were both shown to damage the liver.

All of these examples illustrate the difficulties in detecting harmful effects in advance of actually giving the drug in question to humans. Although the governments of most countries require drugs to be tested on animals before use on humans, these tests can be deceptive. Harmful effects seen in humans are often not seen in rodents, and it is also conceivable that drugs, which are perfectly harmless to humans, could have adverse effects on other species. There are also laws limiting the numbers of animals that can be used, meaning that rare toxic effects that only affect animals with certain conditions are unlikely to be seen until the drug is released for use by humans. The advent of good clinical practice in the wake of the thalidomide tragedy has helped to reduce the effects of these problems, but they are still a hazard that needs careful monitoring.

Different Types of Drug Toxicity

Just as drugs come in many varieties and have varying beneficial and adverse effects, so there are different ways in which they can cause potentially toxic effects. Often it is not the drug itself that causes the trouble, but a breakdown product or 'metabolite', which remains active and has unwanted actions. There are even more complex situations than this, such as in the case of thalidomide, where the drug taken is broken down by the body into two components. One of these acts to reduce morning sickness, while the other causes malformation of the foetus.

In some cases, drugs are in fact chosen for their toxic nature when being used to treat malignant tumours and metastases. In these cases, the cytotoxic and mutagenic properties of these drugs are selected to kill mainly the cancer cells, but can also have negative effects on normal cells.

Cytotoxicity

This is the simplest form of drug toxicity, where the drug or an active metabolite causes serious damage to the cells. Often, the cells of a specific organ are affected, causing a potentially fatal loss of function of the liver or kidney, damage to the eyes or ears, or abnormal clotting of the blood. Other organs can also be affected: aspirin, for example, is known to cause stomach ulcers in some patients, especially after an overdose.

Another well-known cytotoxic drug is paracetamol, also known as acetaminophen, which can fatally damage the liver when taken in high doses. This happens because the body into various metabolites breaks down paracetamol, one of which is NAPQI. This is harmful to the liver, but is normally inactivated by glutathione. However, the body only has limited reserves of glutathione, and an overdose of paracetamol will produce enough NAPQI to deplete these reserves, allowing the NAPQI to destroy the liver cells freely. Paracetamol overdose is treated using methionine or N-acetylcysteine, both of which raise the levels of glutathione and thus inactivate the harmful NAPQI. It is now possible to determine quickly the level of paracetamol in the blood by using anenzyme immunoassay, allowing those with a known toxic level to be treated promptly.

Carcinogenicity

Some drugs can cause damage or mutations to DNA, leading to changes in cell metabolism and the promotion of tumours. Carcinogenic drugs may also activate oncogenes - versions of genes, which promote disregulation of cell processes - leading to the transformation of normal cells into cancerous cells. An example of a drug with carcinogenic properties is stilboestrol, an oestrogen analogue that increases the probability of developing endometrial cancer or breast cancer.

Mutagenicity

Some drugs can cause permanent changes to the DNA of germ cells - egg cells and sperm cells - leading to mutations, which are inherited by a patient's children. One example is nitrogen mustard, a compound similar to mustard gas, which is used to destroy cancer cells in chemotherapy. The drug destroys cancer cells by linking their DNA strands together with a nitrogen atom, but can also link the DNA of germ cells, leading to deletions of DNA bases, thereby causing mutations. For this reason, known mutagens are usually only used deliberately to kill cancer cells.

Teratogenicity

As in the case of thalidomide, some drugs can cause defects in the development of the foetus, leading to gross abnormalities of the baby: the term for this kind of toxicity, teratogenicity, derives from the Greek *teratos*, meaning 'monster'. The type of abnormality seen depends on the stage at which the drug is taken, as different organs develop at different times during pregnancy.

For instance, thalidomide taken after 21 days of pregnancy will often lead to lack of external ears and paralysis of the cranial nerves, whereas thalidomide taken after 27 days will lead to limb malformation. This latter deformity is known as phocomelia, meaning 'seal-like limbs'.

Drug Allergy

Allergies are a well-known problem, with the offending substances ranging from peanuts to pollen. Drug allergies work in the same way, with the body developing antibodies to a drug, causing the immune system to overreact, leading to the usual range of symptoms including itching, wheezing and rashes. However, when the drug causes an anaphylactic shock, where the patient's immune system responds strongly enough to cause difficulties in breathing, low blood pressure, inflammation and even heart failure, the drug is noted as toxic to that individual, and cannot be given to them again. Penicillin is a good example of a drug that can cause allergic reactions and anaphylaxis, with a breakdown product of the drug causing an immune reaction.

Measuring Toxicity

A simple yet crude measure of the toxicity of a drug is the LD_{50}, which is the dose of a drug which kills 50% of treated animals within a specified short period of time. This quantity varies widely from ten grams per kilogram of body weight for alcohol, to just 0.00001 milligrams per kilogram for the botulinus toxin produced by the botulinium bacterium. Various other animal tests are used, and drugs are also added to colonies of bacteria and cultures of human cells to determine their carcinogenic and mutagenic effects.

Avoiding Unwanted Toxicity

Unwanted toxic effects can be avoided by banning dangerous drugs, or by strictly controlling their uses. Drugs must go through both rigorous animal testing and small-scale trials on humans to detect unexpected effects, with some dangerous drugs only being dropped after humans have reacted to them. It is now common practice for pharmaceutical companies to advise doctors not to prescribe their drugs to pregnant women and up to half the drugs listed in the

British National Formulary come with such warnings. Women of child-bearing age are often advised to use contraception while taking potentially teratogenic drugs. Meanwhile, drugs that are potentially fatal after an overdose are sold in restricted amounts in blister packs, making it difficult to take enough pills for the drug to have toxic effects. Childproof containers have also been developed and clear labelling is used to warn patients of risks. Doctors are now very careful when prescribing drugs, which may interact with one another to cause increased effects, and they are also careful to keep up-to-date with the latest toxicity reports in journals. Previously unknown adverse effects of drugs are reported to the authorities using a standard procedure.

Treating Unwanted Toxicity

Treatment of toxicity refers to the treatment of overdoses, where the drug is likely to cause harm if its effects are not quickly countered. The first line of defence is to remove the drug from the patient before it is fully absorbed, and techniques include:

- Irrigation to remove drugs applied to the eyes or skin.
- Gastric lavage, where the stomach is washed out and drained using tubes
- Activated charcoal, which is swallowed and soaks up the drug from the gut.
- Ipecac syrup, which causes vomiting in order to empty the stomach.
- Cathartics, laxatives which purge the drug from the gut.

A second line of defence involves the removal of the drug from the bloodstream by various methods, including:

- Changing urine pH to increase excretion of the drug into the urine.
- Forced diuresis, where drugs are given to increase urine production.
- Haemodialysis, where the blood is passed through a machine to remove the drug.

- Exchange transfusion to replace the patient's drug-filled blood with fresh blood.

A vitally important line of defence is the administration of an antidote, which either removes the harmful substance from the blood or counters its effects. Pralidoxine is used as an antidote to nerve gas, and ethanol is given to counter methanol, while hydroxycobalamin is used to treat cyanide poisoning, although it must be administered quickly. An example of an antidote with an opposing effect is atropine, which in large doses causes muscle paralysis: this is used to oppose the effects of an overdose of anti-cholinesterases, drugs which increase the firing of nerves attached to muscles.

And Finally

Having read the above, it is very easy to worry about the possibility of any drug having potentially harmful effects. It is of course true that many drugs can be dangerous at high doses, but it is for this reason that doctors only prescribe small quantities of drugs. Drugs, which can have carcinogenic effects usually only, cause cancer in a minority of patients and are used because their benefits vastly outweigh the risks. Meanwhile, potentially teratogenic drugs are not prescribed to pregnant women and mutagenic drugs are usually only used in chemotherapy, where the mutagenic effect is required to kill off cancer cells. All in all, doctors are provided with a vast quantity of information about harmful effects in the British National Formulary and have been trained to make careful decisions about which drugs to prescribe to patients.

A new computational method that searches an enormous database of protein structures could allow researchers to predict a drug's potential side effects without breaking out a single test tube. The technique, developed by researchers at the University of California, San Diego (UCSD), could also be applied to existing drugs to explain known side effects or to identify additional uses.

"The approach that they're taking is different and certainly way more computationally intensive than other approaches,"a pharmacologist at the University of North

Carolina, Chapel Hill, who was not involved with the work. Drugs work by latching onto very specific receptor sites on protein targets, much like a key fitting into a lock. If a drug also happens to latch onto another, unintended target, it may produce unwanted side effects. The new method analyses the shape of the intended lock and then combs through the structures of other proteins to look for similarly shaped locks. If such a lock is found, an algorithm rigorously determines whether the drug fits snugly into it. If it does, the technique has probably identified what's called an "off-target."

The researchers, led by UCSD pharmacologist Philip Bourne, didn't stop there. Once they had identified an off-target, they worked backward, looking for other drugs known to bind to it and then using their technique to determine whether those drugs could also bind to the original target. If so, the researchers had further evidence that the two proteins—the intended target and the unintended target—had highly similar drug receptor sites. Then, by assessing the biological function of the off-target, they could postulate possible side effects of the drug being tested.

DRUG SIDE EFFECTS

All medicines may cause side effects, but many people have no, or minor, side effects.Check with your doctor if any of these most COMMON side effects persist or become bothersome when using Methotrexate:

Acne; chills and fever; dizziness; flushing; general body discomfort; hair loss; headache; infertility; irregular periods; itching; loss of appetite; lowered resistance to infection; miscarriage; nausea; sensitivity to sunlight; sore throat; speech impairment; stomach pain; swelling of the breast; unusual tiredness; vaginal discharge; vomiting.

Seek medical attention right away if any of these SEVERE side effects occur when using Methotrexate:

Severe allergic reactions (rash; hives; itching; difficulty breathing; tightness in the chest; swelling of the mouth, face, lips, or tongue); black or bloody stools; blood in the urine; bone pain; calf pain/swelling; change in amount of urine; chest pain;

confusion; dark urine; diarrhea; dry cough; enlarged glands; fatigue; fever or chills; inflammation of the pancreas (stomach tenderness, nausea, vomiting, fever, increased pulse rate); irregular heartbeat; mental changes; mouth sores; muscle weakness; persistent sore throat; red, swollen, or blistered skin; seizures; serious infection (herpes, hepatitis, blood infection); trouble breathing; unusual bleeding or bruising; unusual pain and discoloration of the skin; vision changes; vomit that looks like coffee grounds; yellowing of skin or eyes.

Tablets

All medicines may cause side effects, but many people have no, or minor, side effects.Check with your doctor if any of these most COMMON side effects persist or become bothersome when using Methotrexate Tablets:

Dizziness; headache; loss of appetite; mild hair loss; nausea; stomach pain or upset; tiredness; vomiting.Seek medical attention right away if any of these SEVERE side effects occur when using Methotrexate Tablets:

Severe allergic reactions (rash; hives; itching; difficulty breathing; tightness in the chest; swelling of the mouth, face, lips, or tongue); black, tarry stools; blood in the urine; calf or leg pain or swelling; change in the amount of urine; chest pain; coma; confusion; coughing up blood; dark urine; diarrhea; difficult or painful urination; dry cough; fever, chills, or sore throat; menstrual changes; mental or mood changes; mouth sores; one-sided weakness; red, swollen, peeling, or blistered skin; seizures; severe or persistent nausea or vomiting; shortness of breath; speech changes; unusual bleeding or bruising; unusual pain and discoloration of the skin; unusual tiredness or weakness; vision changes or vision loss; yellowing of skin or eyes.

Methotrexate

In General, The Incidence And Severity Of Acute Side Effects Are Related To Dose And Frequency Of Administration. The Most Serious Reactions Are Discussed Above Under Organ System Toxicity In The Precaution Section. That Section Should

Also Be Consulted When Looking For Information About Adverse Reactions With Methotrexate.

The most frequently reported adverse reactions include ulcerative stomatitis, leukopenia, nausea, and abdominal distress. Other frequently reported adverse effects are malaise, undue fatigue, chills and fever, dizziness and decreased resistance to infection.

Other adverse reactions that have been reported with Methotrexate are listed below by organ system. In the oncology setting, concomitant treatment and the underlying disease make specific attribution of a reaction to Methotrexate difficult.

- *Alimentary System:* Gingivitis, pharyngitis, stomatitis, anorexia, nausea, vomiting, diarrhea, hematemesis, melena, gastrointestinal ulceration and bleeding, enteritis, pancreatitis.
- *Blood and Lymphatic System Disorders:* Suppressed hematopoiesis causing anemia, aplastic anemia, pancytopenia, leukopenia, neutropenia, and/or thrombocytopenia, lymphadenopathy and lymphoproliferative disorders (including reversible). Hypogammaglobulinemia has been reported rarely.
- *Cardiovascular:* Pericarditis, pericardial effusion, hypotension, and thromboembolic events (including arterial thrombosis, cerebral thrombosis, deep vein thrombosis, retinal vein thrombosis,thrombophlebitis, and pulmonary embolus).
- *Central Nervous System:* Headaches, drowsiness, blurred vision, transient blindness, speech impairment including dysarthria and aphasia, hemiparesis, paresis and convulsions have also occurred following administration of Methotrexate. Following low doses, there have been occasional reports of transient subtle cognitive dysfunction, mood alteration, unusual cranial sensations, leukoencephalopathy, or encephalopathy.
- *Hepatobiliary Disorders:* Hepatotoxicity, acute hepatitis, chronic fibrosis and cirrhosis, decrease in serum albumin, liver enzyme elevations.

- *Infection:* There have been case reports of sometimes fatal opportunistic infections in patients receiving Methotrexate therapy for neoplastic and non-neoplastic diseases. Pneumocystis carinii pneumonia was the most common opportunistic infection. There have also been reports of infections, pneumonia, sepsis, nocardiosis, histoplasmosis, cryptococcosis, Herpes zoster, H. simplex hepatitis, and disseminated H. simplex.
- *Musculoskeletal System:* Stress fracture.
- *Ophthalmic:* Conjunctivitis, serious visual changes of unknown etiology.
- *Pulmonary System:* Respiratory fibrosis, respiratory failure, interstitial pneumonitis deaths have been reported, and chronic interstitial obstructive pulmonary disease has occasionally occurred.
- *Skin:* Erythematous rashes, pruritus, urticaria, photosensitivity, pigmentary changes, alopecia, ecchymosis, telangiectasia, acne, furunculosis, erythema multiforme, toxic epidermal necrolysis, Stevens-Johnson Syndrome, skin necrosis, skin ulceration, and exfoliative dermatitis.
- *Urogenital System:* Severe nephropathy or renal failure, azotemia, cystitis, hematuria; defective oogenesis or spermatogenesis, transient oligospermia, menstrual dysfunction, vaginal discharge, and gynecomastia; infertility, abortion, fetal defects.

Other rarer reactions related to or attributed to the use of Methotrexate such as nodulosis, vasculitis, arthralgia/myalgia, loss of libido/impotence, diabetes, osteoporosis, sudden death, reversible lymphomas, tumor lysis syndrome, soft tissue necrosis and osteonecrosis. Anaphylactoid reactions have been reported.

Double-Blind Rheumatoid Arthritis Studies

The approximate incidences of Methotrexate attributed (*i.e.*, placebo rate subtracted) adverse reactions in 12 to 18 week double-blind studies of patients (n = 128) with rheumatoid arthritis treated with low-dose oral (7.5 to 15 mg/week) pulse Methotrexate are listed below. Virtually all of these patients were on concomitant nonsteroidal anti-inflammatory drugs and some were also taking low dosages of corticosteroids. Hepatic histology was not examined in these short-term studies.

Incidence Greater Than 10%: Elevated liver function tests 15%, nausea/vomiting 10%. Incidence 3% to 10%: Stomatitis, thrombocytopenia, (platelet count less than 100,000/mm3). Incidence 1% to 3%: Rash/pruritus/dermatitis, diarrhea, alopecia, leukopenia (WBC less than 3000/mm3), pancytopenia, dizziness.

Two other controlled trials of patients (n = 680) with Rheumatoid Arthritis on 7.5 mg to 15 mg/wk oral doses showed an incidence of interstitial pneumonitis of 1%. Other less common reactions included decreased hematocrit, headache, upper respiratory infection, anorexia, arthralgias, chest pain, coughing, dysuria, eye discomfort, epistaxis, fever, infection, sweating, tinnitus, and vaginal discharge.

Adverse Reactions in Psoriasis

There are no recent placebo-controlled trials in patients with psoriasis. There are two literature reports describing large series (n = 204, 248) of psoriasis patients treated with Methotrexate. Dosages ranged up to 25 mg per week and treatment was administered for up to four years. With the exception of alopecia, photosensitivity, and "burning of skin lesions" (each 3% to 10%), the adverse reaction rates in these reports were very similar to those in the rheumatoid arthritis studies. Rarely, painful plaque erosions may appear.

Adverse Reactions in JRA Studies

The approximate incidences of adverse reactions reported in pediatric patients with JRA treated with oral, weekly doses of Methotrexate (5 to 20 mg/m2/wk or 0.1 to 0.65 mg/kg/wk) were as follows (virtually all patients were receiving concomitant nonsteroidal anti-inflammatory drugs, and some also were taking low doses of corticosteroids): elevated liver function tests, 14%; gastrointestinal reactions (*e.g.*, nausea, vomiting, diarrhea), 11%; stomatitis, 2%; leukopenia, 2%; headache, 1.2%; alopecia, 0.5%; dizziness, 0.2%; and rash, 0.2%. Although there is experience with dosing up to 30 mg/m2/wk in JRA, the published data for doses above 20 mg/m2/wk are too limited to provide reliable estimates of adverse reaction rates.

SIDE EFFECTS BY BODY SYSTEM

Hematologic

Hematologic side effects have included myelosuppression, which is one of the primary toxic effects of methotrexate. Methotrexate suppressed hematopoiesis has been reported to have caused anemia, aplastic anemia, pancytopenia, leukopenia, neutropenia, thrombocytopenia, lymphadenopathy, and lymphoproliferative disorders including reversible hypogammaglobulinemia (which has been reported rarely).

Preexisting myelosuppression or low hematologic cell counts are contraindications to the use of this drug, particularly in patients with rheumatoid arthritis or psoriasis. Close monitoring of the CBC is mandatory. Profound count nadirs may require therapy discontinuation, at least temporarily. Folate therapy and/or leucovorin rescue may be preventive or palliative. Treated patients who become febrile should be assumed to have neutropenia until proven otherwise.

Cytopenia occurs in 5% to 25% of patients with rheumatoid arthritis (RA) who receive long-term therapy. Risk factors include renal dysfunction, preexisting folate deficiency, increased mean corpuscular volume value, advanced age, concomitant use of other anti-folate medications (such as trimethoprim-sulfamethoxazole), and possibly hypoalbuminemia, concomitant infection, history of bone marrow injury, surgery, and concurrent use of NSAIDs or probenecid. Pancytopenia is rarely observed in patients with rheumatoid arthritis. Bone marrow recovery typically occurs within two weeks after the withdrawal of MTX.

Gastrointestinal

Gastrointestinal side effects are usually controlled by folate supplementation (1 to 5 mg orally, given 4 hours before MTX or 1 mg orally once a day if on low dose therapy, as in rheumatoid arthritis), dosage reduction, dividing the dose over a 12 to 24 hour period, withholding the drug, or giving it parenterally.

Extremely rare cases of colitis and toxic megacolon have been associated with the use of MTX. Gastrointestinal side effects, especially with high-dose administration, may be expected. Serious nausea, vomiting, diarrhea, or stomatitis (10% to 80% of patients followed in long-term studies) may result in symptomatic dehydration. Other frequently reported GI side effects, particularly after high-dose therapy, include gingivitis, pharyngitis, stomatitis, anorexia, hematemesis, melena, gastrointestinal ulceration and bleeding, and enteritis. GI symptoms are often eliminated by folate supplementation. Folate does not affect efficacy of MTX.

Hepatic

Hepatic side effects including hepatotoxicity, acute hepatitis, chronic fibrosis and cirrhosis, decrease in serum albumin, and liver enzyme elevations have been reported.

Methotrexate can cause acute elevations of liver function tests (elevated serum transaminases in 15% of patients with rheumatoid arthritis (RA) on low-dose therapy) or chronic hepatotoxicity (fibrosis and cirrhosis). The incidence of liver fibrosis and cirrhosis in patients with RA (low doses) averages 3% to 7% and 0.1%, respectively.

Meta-analysis has revealed the incidence of progression of liver disease (worsening of 1 grade on the histologic classification of Roenigk) in patients with RA or psoriatic arthritis averages 27%, or 7% per gram of MTX (total dose) given. Chronic hepatotoxicity typically develops only after chronic use of higher doses (2 years or more of total doses of 1.5 grams or more), is more likely in patients who ingest ethanol, who are aged, who are obese, who have chronic renal insufficiency, or who have diabetes.

In general, the following serve as guidelines for patients with non-oncologic conditions:

- Screen all patients to be treated with complete liver blood tests (transaminases, albumin, alkaline phosphatase, bilirubin), hepatitis B and C serologic studies (some experts recommend viral serology's only in patients with abnormal liver function tests), baseline serum creatinine and complete blood count (CBC);
- Baseline liver biopsy if the patient has preexisting liver disease, persistently abnormal baseline AST (aspartate aminotransferase), history of excessive alcohol consumption (greater than 3 drinks/day), or chronic hepatitis B or C infection;
- Monitor hepatic enzymes every 4 to 8 weeks (with the understanding that they are not necessarily predictive of fibrosis/cirrhosis), discontinuing MTX if serum transaminase levels meet or exceed 2 to 3x baseline (some guidelines recommend withholding MTX for 1 to 2 weeks if significant abnormalities persist);
- Liver biopsy should be considered if significant hepatic enzyme abnormalities persist for 2 to 3 months. Liver biopsy should be performed if 6 of 12 monthly AST levels are above the upper limit of normal or serum albumin levels are less than normal in the setting of well controlled disease;
- If liver biopsy shows Roenigk grade I, II, or IIIA (mild fibrosis), MTX may be resumed with monitoring as described above; if liver biopsy show Roenigk grade IIIB (moderate fibrosis) or IV (cirrhosis), MTX should be discontinued;
- MTX should be discontinued in patients with significant liver abnormalities, as described above, who refuse liver biopsy;
- It is still debated whether post-therapy biopsy should be performed every 2 to 3 years or after each 1.5 to 2.0 grams cumulative dose given, regardless of liver chemistries.

*Pre-biopsy coagulation studies and withholding of aspirin and other NSAIDs are recommended.

Immunologic

New and/or opportunistic infection can arise during or after therapy with methotrexate due to drug-induced immune suppression. Infections during MTX therapy may occur in up to 58% of patients during low-dose therapy (as in RA). Patients are at great risk after high-dose therapy. MTX should be held prior to and during elective surgery to minimize the risk of infectious complications.

Limited data have shown that the use of low-dose MTX for patients with rheumatoid arthritis is associated with decreased CD8 and naive CD4 T lymphocytes after only 8 weeks of therapy.Methotrexate is usually contraindicated in patients with overt or laboratory evidence of immunodeficiency syndromes. Immunization may be ineffective when given during methotrexate therapy. Immunization with live virus vaccines is generally not recommended. There have been reports of disseminated vaccinia infections after smallpox immunization in patients receiving methotrexate therapy.

Immunologic side effects including case reports of sometimes-fatal opportunistic infections have been reported. Pneumocystis carinii pneumonia has been reported most frequently. Also of concern are infections associated with severe immunosuppression, such as disseminated herpes zoster, Listeria meningitis, Mycobacterium avium intracellulare pneumonia, and systemic fungal infections (cryptococcosis, nocardiosis, aspergillosis, and histoplasmosis).

Nervous system

Nervous system side effects include headaches, dizziness, drowsiness, blurred vision, subtle cognitive dysfunction, moodiness, tinnitus or unusual cranial sensations. After intrathecal administration, headache, back pain, fever, and even transient paraplegias have been reported. A case of acute chorea and a case of a woman who couldn't speak after receiving methotrexate by intrathecal administration have also been reported. Serious neurotoxicity has been associated with the use of high-dose MTX after intrathecal or intraventricular administration to patients who have undergone craniospinal irradiation, but has also been described in other patients who have received low-dose oral therapy.The development of serious neurotoxicity is more likely if MTX is given in high doses to patients who have undergone craniospinal irradiation.

Significant neurotoxicity can present as an acute stroke-like encephalopathy or a chronic leukoencephalopathy. Unique features of the former include its acute onset (seizures, confusion, hemiparesis, speech problems, loss of consciousness) and reversibility within days.The use of high-dose IV MTX has been associated with the development of chronic delayed leukoencephalopathy in patients with or without a history of craniospinal irradiation. A syndrome of subtle personality changes followed by a progressive dementia, focal seizures, pseudobulbar palsy, spastic quadriparesis, and stupor may begin several months after the initiation of therapy. Affected patients often improve after discontinuation of therapy.

Intrathecal or intraventricular administration of MTX has been associated with the acute onset of headache, nausea, vomiting, fever, back pain, dizziness, and meningismus. Acute symptoms usually resolve within one to two days. In some cases, subacute neurotoxicity, with myelopathy or encephalopathy, may occur within days to weeks. Subacute neurotoxicity is usually characterized by paresis/plegia, dementia, confusion, tremor, ataxia, irritability, and somnolence. Delayed leukoencephalopathy, as described above, has also been associated with the use of intrathecal MTX. The condition can improve but can be progressive and fatal.Progressive dementia and leukoencephalopathy has been reported in at least one case after low-dose, oral administration of MTX for the treatment of rheumatoid arthritis.

Respiratory

Two patterns of pulmonary toxicity have been reported a

hypersensitivity reaction and a toxic reaction with diffuse alveolar damage and nonspecific lung injury. Risk factors are unknown, but may include male gender, cigarette smoking, and concomitant use of an NSAID.

The hypersensitivity reaction is characterized by interstitial pneumonitis, granuloma formation, and the development of bronchopneumonia. Some patients have asthmatic symptoms.

Interstitial pneumonitis may be associated with duration of treatment or weekly or cumulative dose. The prevalence of this complication ranges from 2.5% to 7.5% in patients with psoriatic or rheumatoid arthritis. One prospective study has shown that the cumulative dose of MTX is independently related to small, but significant, increases in residual volume and that MTX may cause mild degrees of air trapping after long-term therapy.

The differential diagnosis of MTX pneumonitis includes infectious pneumonia, hypersensitivity pneumonitis caused by other drugs and autoimmune alveolitis due to rheumatoid arthritis (RA). Pneumocystis carinii pneumonia (PCP) can resemble MTX pneumonitis, and bronchoalveolar lavage (BAL) is a useful tool to distinguish between the two. BAL is not necessarily helpful in distinguishing MTX pneumonitis from RA alveolitis, however, since an elevated lymphocyte count and an increased CD4/CD8 ratio are observed in both conditions.

Pulmonary function tests in affected patients reflect a restrictive ventilatory defect and decreased oxygen-diffusing capacities. A mild peripheral eosinophilia is often present.

Respiratory side effects including toxicity can occur at any dosage in 3% to 11% of patients, and can mimic infectious pneumonia. Interstitial pneumonitis has emerged as one of the most unpredictable and potentially life-threatening side effects from low-dose MTX therapy. It presents as fever, dry/nonproductive cough, dyspnea, hypoxemia, and/or epistaxis, usually with chest X-ray infiltration. Signs or symptoms of infection must be treated seriously due to the possibility of MTX-induced neutropenia.

Renal

Renal side effects include renal insufficiency, which is most commonly associated with high-dose MTX since, after these doses, the concentration of a major circulating metabolite, 7-OH MTX, can precipitate in the renal tubule. Concomitant use of other potentially nephrotoxic drugs, including NSAIDs, and preexisting renal insufficiency are risk factors. There is a higher risk of nephrotoxicity in patients with underlying renal dysfunction. Aggressive and adequate hydration and urinary alkalinization helps minimize the risk of MTX-induced nephropathy, cystitis, and hematuria. If monitoring reveals renal dysfunction, decreasing MTX doses or discontinuing the drug altogether may improve renal function.

General

General side effects, which have been reported frequently, include malaise, fatigue, and chills. Less commonly reported effects include arthralgias and myalgias. Leucovorin is typically given to diminish toxicity and counteract the adverse effects of high-dose therapy.

There is an increased risk of methotrexate (MTX) toxicity in patients who have "third space" accumulations of fluid, such as pleural effusions. MTX accumulates in these spaces and becomes cleared from these spaces slowly. These "deep pools" of MTX markedly increase the risk of toxicity, particularly gastrointestinal toxicity (mucositis) and should be evacuated before therapy begins.

Genitourinary

Genitourinary side effects may seriously affect either sex. Women may experience menstrual dysfunction, vaginal discharge, abortion, or infertility. Decreased libido has been described in each sex. Defective oogenesis and spermatogenesis is usually transient.

Dermatologic

Cases of severe, sometimes fatal, dermatologic reactions, such as toxic epidermal necrolysis, Stevens-Johnson syndrome, exfoliative dermatitis, skin necrosis, and erythema multiforme, have been reported in children and adults, within days of oral, intramuscular, intravenous, or intrathecal methotrexate administration.

Reactions have been observed after single or multiple, low, intermediate or high doses of methotrexate for the treatment of neoplastic and non-neoplastic diseases. Alopecia typically resolves several months after discontinuation.

Dermatologic side effects include erythematous rashes, desquamation, epidermal necrosis, pruritus, urticaria, photosensitivity, pigmentary changes, alopecia, ecchymosis, telangiectasia, acne, cutaneous vasculitis, furunculosis, and alopecia. Painful plaque erosions have rarely been reported (when MTX was used to treat psoriasis).

Musculoskeletal

Musculoskeletal side effects from low-dose MTX in patients with rheumatoid arthritis include rare instances of accelerated nodulosis. This problem is more commonly associated with MTX than other second-line agents. The use of MTX may be associated with decreased bone density in a syndrome called "MTX osteopathy" (this potential complication requires further investigation). This syndrome may be especially important in postmenopausal women.

Hypersensitivity

Hypersensitivity side effects including anaphylaxis have been reported rarely.

Oncologic

Oncologic side effects include some evidence that methotrexate may be oncogenic, particularly with respect to the development of some lymphomas and leukemias.

Large retrospective studies have shown that hematologic malignancies are uncommon in patients with RA treated with disease-modifying antirheumatic drugs, including MTX. These studies have shown that there does not appear to be a relationship between the peak or cumulative dose of the duration of MTX therapy and the subsequent development of hematologic malignancy. The histologic types of hematologic malignancies seen in MTX-treated patients do not appear different from those seen in patients with RA treated with other disease-modifying antirheumatic drugs. Underlying rheumatoid arthritis (RA) or Sjogren's syndrome are independent risk factors for the development of non-Hodgkin's lymphoma.

Cardiovascular

Cardiovascular side effects including pericarditis, pericardial effusion, myocardial ischemia, hypotension and ventricular arrhythmias have rarely been associated with MTX. Thromboembolic events including arterial thrombosis, cerebral thrombosis, deep vein thrombosis, retinal vein thrombosis, thrombophlebitis, and pulmonary embolus have also been reported with methotrexate use. Chemical pleuritis secondary to MTX has been identified as the cause of chest pain in some patients after high-dose therapy.

Endocrine

Endocrinologic side effects have included gynecomastia associated with the use of low-dose MTX in patients with rheumatoid arthritis.

Ocular

Ocular side effects reported with methotrexate include conjunctivitis and serious visual changes of unknown origin.

4

Chemical Metabolism

MODIFICATION PATHWAY

The evolutionarily conserved phenomenon RNA interference (RNAi), the process by which specific mRNAs are targeted for degradation by complementary short-interfering RNAs (siRNAs), has increasingly become a powerful tool for genetic analysis and is likely to become a potent therapeutic approach for gene silencing. Consequently, understanding the mechanism of RNAi has become critical for developing the most effective RNAi methodologies for both laboratory and clinical applications. The general mechanism of RNAi involves the cleavage of double-stranded RNA (dsRNA) to short 21–23-nt siRNAs. This processing event is catalyzed by Dicer, a highly conserved, dsRNA-specific endonuclease that is a member of the RNase III family.

Processing by Dicer results in siRNA duplexes that have 52 -phosphate and 32 -hydroxyl termini, and subsequently, these siRNAs are recognized by the RNA-induced silencing complex. Active RISC complexes (RISC*) promote the unwinding of the siRNA through an ATP-dependent process, and the unwound antisense strand guides RISC* to the complementary mRNA.The targeted mRNA is then cleaved by RISC* at a single site that is defined with regard to where the 52 -end of the antisense strand is bound to the mRNA target sequence. For RNAi-mediated mRNA cleavage and degradation to be successful, 52 -phosphorylation of the antisense strand must occur, and the double helix of the antisense-target mRNA duplex must be in the A form.

One highlighted difference between mammalian RNAi and RNAi in other eukaryotes is the lack of an amplification system for long-term persistence of RNAi in mammalian cells. For example, in Drosophila, ~35 molecules of dsRNA can silence ~1000 copies of the targeted mRNA per cell and can persist over the course of many generations. In mammalian cells, RNAi only persists effectively for an average of ~66 h before the siRNA is likely diluted out over the course of several cell divisions. The amplification that is seen in flies and other lower eukaryotes can potentially be attributed to three factors. One is that the conversion of long trigger dsRNA to smaller 21–23-nt siRNAs by Dicer adds a degree of RNAi amplification, whereas in mammalian cells long trigger dsRNA invokes the interferon response that activates the protein kinase PKR.

This suggests that only siRNA transfections successfully trigger RNAi in mammalian cells without other side effects, and thus, no amplification would take place through the processing of longer RNAs. A second factor in amplification is the presence of RNA-dependent RNA polymerase (RdRP), which has been found in plants, worms, fungi, and flies. RdRP has been postulated to amplify target mRNA, through a random, degradative PCR model, into dsRNA, which can be targeted by Dicer. However, no RdRP homologs have been found in mammalian cells, and the 32 -OH that is required for RdRP-dependent degradative PCR is not required for RNAi in mammalian cells, indicating that PCR-based amplification likely does not occur in mammals.

A third factor in amplification may be the high enzymatic turnover rate of RISC* during the targeting and cleavage of mRNA, which may add a degree of amplification to RNAi induction in all eukaryotes, including mammals. However, as the persistence of RNAi occurs for only a short period of time, finding methods for increasing the longevity of siRNAs in human cells will be fundamental for applying RNAi to laboratory and therapeutic applications.

To address this issue of siRNA stability for prolonging the duration of dsRNA-mediated gene silencing and to further dissect the mechanism of RNAi in human cells, various chemically modified nucleotides predicted to affect siRNA stability were incorporated into siRNAs to study whether specific modifications increased or decreased the efficacy and persistence of RNAi in vivo. The most important of these modifications was to the 22 -OH of the ribonucleotide that distinguishes RNA from DNA and is required for the nucleophilic attack occurring during the hydrolysis of the RNA backbone, the reaction catalyzed by degradative RNases.

Our results showed that the 22 -OH was not required for RNAi, indicating that structural rather than chemical properties of siRNA–mRNA duplexes were the key to inducing RNAi and that RISC* did not require the 22 -OH for ribonuclease activity. 22 -modified siRNAs also increased the persistence of RNAi in human cells. Modifications that stabilized base-pairing interactions were also incorporated into the antisense strand of siRNAs and were able to initiate RNAi, signifying that this class of chemical modifications could be used to increase the targeting efficiency of siRNAs for mRNA target sequences and for allele-specific inhibition of gene expression.

Other chemical modifications affected the formation of the major groove of the A-form helix of the antisense-siRNA–target-mRNA duplex, and potentially disrupted H-bonds or sterically hindered protein contacts, most probably preventing the RISC* complex from stably interacting with the dsRNA duplex. These modifications completely abolished RNAi, demonstrating that an intact major groove in the A-form helix and stable RNA–protein interactions were required for RNAi in human cells. Finally, previous observations of psorelan cross-linked siRNAs implied that unwinding of siRNA occurred from the 52 -end of the antisense strand and that complete unwinding may not be necessary for effective RNAi.By using mismatched or chemically modified nucleotides on either the 32 or 52 half of the antisense strand within the siRNA duplex, we have shown here that RNAi depended on the integrity of the 52, and not the 32, half of the siRNA duplex, as defined by the antisense strand. Altogether, these results gave insight into the essential biochemical properties of functional siRNAs and how specific changes in the siRNA structure can affect the efficacy of RNAi. Furthermore, these studies present new methodologies for improving the stability and utility of siRNAs for future RNAi applications.

BETA-BLOCKERS

Beta blockers, also known as beta-adrenergic blocking agents, are medications that reduce your blood pressure. Beta blockers work by blocking the effects of the hormone epinephrine, also known as adrenaline. When you take beta blockers, the heart beats more slowly and with less force, thereby reducing blood pressure. Beta blockers also help blood vessels open up to improve blood flow.

Beta blockers have been called "the musicians underground drug." Often musicians form their opinions, and may risk their health, based on locker-room-type information.

Performance anxiety can be a deeply personal subject for musicians, and many are reluctant to discuss all the possible remedies. It is our intention to bring this subject into the open, and to provide accurate information to inform personal opinions and decisions.

BETA BLOCKERS (SUCH AS INDERAL)

Beta blockers block the receptors for the physical effects of a person's natural fight or flight response. They are not sedatives, and they can't help anxiety of a purely psychological nature. Beta receptors are found in a number of places in the body: heart, lung, arteries, brain and uterus, to name a few. Like a key in a lock, beta blockers chemically fit into beta receptors and prevent norepinephrine from binding to the receptors that cause the symptoms of the fight-or-flight response.

The degree of these effects depends on the dose and the individual's sensitivity to the medication. Peak effect occurs in one to one and a half hours. Ideally, this could allow a performer to play at his or her best, without the distraction or interference of excessive fight or flight symptoms. Blocking beta receptors can cause decreased heart rate; decreased force of heart contractions; bronchoconstriction (can cause asthma attacks in people with asthma); uterine contractions; decreased blood pressure; relief of migraines; and decreased tremor.

The beta receptors found in the different areas of the body are not all the same, thus different beta blockers may affect these areas differently. For instance, metoprolol (Lopressor) and atenolol (Tenormin) are beta-1 selective, which means they block only beta-1 receptors found primarily in the heart, but not the beta-2 receptors found in the lung and uterus. Thus, they can decrease blood pressure, heart rate and force of cardiac contraction, but are less likely to cause bronchoconstriction and uterine contractions. This selectivity is not absolute and depends on the dose.

Some beta blockers enter the brain better than others. Propranolol (Inderal) crosses the blood-brain barrier particularly well. This may be why propranolol causes more central nervous system side effects, such as hallucinations, nightmares, and depression, than the beta blockers that do not cross into the brain as easily.

Interestingly, the ability of beta blockers to help anxiety seems related only to their blockade of beta receptors cutside of the brain. Beta blockers will not help the emotional symptoms of stage fright (for example, sleep problems or negative inner voices).

Everyone's body responds differently to beta blockers. For this reason it is important that each person experiment, under medical supervision, to find the ideal dosage for a performance situation. Similarly, it is important to take the beta blocker a few times and under different circumstances before a major performance to be sure the individual has no adverse reactions.

DOCTOR ADIVISE

Everyone, without exception, must obtain their beta blockers from a physician. It is a violation of federal law in the country, and many other countries, to give prescription medications to someone other than the person for whom they were prescribed. Beta blockers are prescription medications for good reason. There are several conditions, such as slow heart rate, bronchial asthma, and heart failure, which can cause serious medical problems, even death, when taking this medication.

In addition, you should know that some psychiatrists and psychologists feel that beta blockers ought only to be used in the context of psychological intervention, and only as a temporary measure.

PROBLEMS ARE BETA BLOCKER

One reference lists 29 different uses, including high blood pressure, angina, irregular heart rhythms, migraines, prevention of a second heart attack, tremors, alcohol withdrawal, anxiety and glaucoma.

In general, a physician can prescribe medications for uses not approved by the Food and Drug Administration, so long as there is scientific literature to support the use. Usually the pharmaceutical company involved will only research and develop a few uses to submit to the FDA to get the medication approved for marketing.

Once the medication wins FDA approval, the pharmaceutical company is only allowed to advertise it for FDA-approved uses. Over time, however, researchers and physicians gain experience using the medication for uses not approved by the FDA. Beta blockers are not FDA approved for use in anxiety or stage fright, but these medications can be prescribed for this purpose because there is scientific literature to support this use.

SAFETY OF BETA BLOCKERS

Only your physician can tell you whether beta blockers are safe for you.

Beta blockers can significantly worsen some medical conditions. As a general rule, beta blockers should not be used in asthmatic people. Beta blockers can worsen congestive heart failure, Raynaud's syndrome, and diabetes. Musicians who take beta blockers often do so in small doses and only on special occasions, so they may experience no side effects at all from their use. However, many side effects have been reported from these medications, including rash, anaphylactic shock (sudden unconsciousness or death), slowed heart rate, low blood pressure, cold extremities, worsening of heart failure, fainting, rapid heart beats, dizziness, fatigue, headache, depression, sleep disturbances, nightmares, hallucinations, short term memory loss, high or low blood sugar, stomach ache, flatulence, constipation, nausea, diarrhea, dry mouth, vomiting, heartburn, bloating, impotence or decreased libido, difficulty urinating, bronchospasm, cough, wheezes, naal stuffiness, joint pain, and muscle cramps.

BETA BLOCKERS CAUSE ADDICTION

Beta blockers do not cause addiction which means physical and psychological dependence on a substance that is beyond the user's control. But beta blockers, when used on a regular basis, can cause physical dependence.

Physical dependence is when the body becomes accustomed to having a medication in the body, thus over time, more medication is required to achieve the same effect. It also means that if the medication is suddenly stopped, the person can experience a rebound in the symptom the medication was being used to treat.

Not all medications cause physical dependence. To develop physical dependence, the medication must be used on a chronic basis.

In the case of beta blockers, physical dependence is not usually an issue when they are used for stage fright, as they are used in low doses and are not taken on a chronic basis. When beta blockers are taken on a chronic basis, as for high blood pressure, suddenly stopping them can cause a severe, dangerous elevation in blood pressure that could lead to a stroke or heart attack.

Addiction causes the person who drinks alcohol, abuses a medication, or takes an illegal substance to exhibit substance-seeking behaviour, perhaps even committing criminal acts to obtain it.

What makes this issue confusing is that alcohol, narcotics, cocaine, and amphetamines can cause both physical dependence and addiction, and usually both must be treated at the same time. Beta blockers do cause physical dependence when used on a chronic basis, but it would be highly unusual for them to cause addiction.

BETA BLOCKERS HELP IN PERFORMANCE

The answer varies greatly among individuals. Obviously, if you have a medical condition that makes beta blockers dangerous for you, they will not help.

Beta blockers don't make you play better by themselves; they just relieve physical problems resulting from the fight or flight response. On the other hand, some musicians feel that adrenalin helps their performing, giving them an edge that adds intensity to the performance.

Beta blockers have not been shown to directly improve a musician's emotional state, except to the extent that some musicians feel better when their physical problems are relieved. If your performance anxiety shows itself mainly in psychological ways (*e.g.* negative inner voices), beta blockers will not help you. The scientific studies and articles we looked at clearly show that beta blockers significantly reduce symptoms that can hinder some people's playing. The musicians in the studies said they felt better about their performance after taking beta blockers, and music critics consistently judged their performances to be better.

A note about dry mouth: Anecdotal reports show that some musicians who've tried beta blockers say they don't help dry mouth, or can even make it worse. In our research, however, we found a study that conclusively showed beta blockers to help dry mouth among brass players.

BETA BLOCKERS FOR PERFORMANCE ANXIETY

As you might expect, opinions vary widely. Here are what a few well-known musicians have said on the subject: Trevor Wye, flutist and teacher: After saying that beta blockers "should seriously be considered for important occasions," he writes: "Readers who may have some misgivings about the use of alcohol or drugs may care to reflect on the effect of nerves and strain on the body which can cause a great deal of damage over a period of time.

However, one should use caution by understanding the side effects of a given drug and by realizing that an addiction to drugs may occur and that this may therefore not be the ideal approach to the problem of stage fright.

DEALING WITH PERFORMANCE ANXIETY

Because the fight or flight response occurs only when we perceive danger, it can be avoided or minimized if we can convince ourselves there is nothing to fear. This is the goal of many non-medication approaches. The physical sensations of fear can also be managed in specific, non-medication ways.

Cognitive therapy, the area of psychotherapy that holds that your understanding of a situation can change your emotional response to it, is often sought by people with stage fright. Hypnosis may also used. If you decide to seek a therapist, try to find one who is experienced in treating anxiety disorders, and especially one who has experience treating performers.

DIFFERENT CLASSES OF BETA-BLOCKERS AND SPECIFIC DRUGS

Beta-blockers that are used clinically can be divided into two classes:

1. Non-selective blockers (block both β_1 and β_2 receptors), or
2. Relatively selective β_1 blockers ("cardioselective" beta-blockers).

Some beta-blockers have additional mechanisms besides beta-blockade that contribute to their unique pharmacologic profile. The two classes of beta-blockers along with specific compounds are listed in the following table. The clinical uses indicated in the table represent both on and off-label uses of beta-blockers. For example, a given beta-blocker may only be approved by the FDA for treatment of hypertension; however, physicians sometimes elect to prescribe the drug for angina because of the class-action benefit that beta-blockers have for angina.

	Clinical Uses					
Class/Drug	**HTN**	**Angina**	**Arrhy**	**MI**	**CHF**	**Comments**
Non-selective β_1/β_2						
Ccarteolol	X					ISA; long acting; also used for glaucoma
Carvedilol	X				X	α-blocking activity
Labetalol	X	X				ISA; α-blocking activity
Nadolol	X	X	X	X		long acting
Penbutolol	X	X				ISA
Pindolol	X	X				ISA; MSA
Propranolol	X	X	X	X		MSA; prototypical beta-blocker
Sotalol			X			several other significant mechanisms
Timolol	X	X	X	X		primarily used for glaucoma
β_1-selective						
Acebutolol	X	X	X			ISA
Atenolol	X	X	X	X		
Betaxolol	X	X	X			MSA
Bisoprolol	X	X	X			
Esmolol	X		X			ultra short acting; intra or postoperative HTN
Metoprolol	X	X	X	X	X	MSA

CLINICAL DEVELOPMENT

The key to the IRCT approach is our ability to combine clinical trials conducted by Encode, our wholly-owned clinical trials subsidiary, with our population data. Encode has enrolled over 3,000 patients in over 30 clinical trials. In an IRCT the trial cohort is enriched by using population-based epidemiological data as well as data on genetic markers, gene-expression patterns and serum biomarkers correlated with a given disease. This enables deCODE's clinical team to select for a trial patients who are susceptible to disease through the pathway targeted by the drug and are thus likely to respond.

Because of the much larger amount of information going into and coming out of an IRCT, we can test therapeutics for major indications more quickly and with significantly smaller cohorts than would normally be required. The results of an IRCT can then be applied to understand and predict the responsiveness of the patient population at large, information that can be used in further clinical development or to bring the drug to market.

As profit and regulatory compliance pressures increase, life science companies are desperately seeking ways to decrease the cost and manage the risk of their clinical trials. The *Clinical Development Strategy and Technology* service provides life science companies with the necessary insight to increase their operational focus and sharpen their study planning and execution skills.

This Health Industry Insights service provides a forward-looking analysis of IT technologies on the critical path and how they are being adopted all along the clinical development value chain. In addition to qualitative assessment, this service includes quantitative market sizing and forecasts to provide a complete picture of the processes and technologies that impact clinical development.

APPROACH

This service develops comprehensive data and unique analysis through focused topical surveys, secondary research, and insights from industry experts, practitioners, and vendors. To ensure relevance, Health Industry Insights' analysts work with subscribers to identify and prioritize specific topics to be covered in research reports.

TOPICS ADDRESSED

Throughout the year, this service will address the following topics:

- The convergence of the eClinical ecosystem
- Clinical performance management and metrics; Clinical data management, Clinical trial management
- Critical tools and systems, including EDC, ePRO, CTMS, CDMS, IVR/IXR
- Evolution towards adaptive trials
- Achieving innovation and optimization of clinical trials IT infrastructure
- Cost containment and risk assessment in clinical development
- eClinical technology landscape, including integration with electronic health record (EHR) and other healthcare IT systems
- Pharmacovigilance and drug safety
- Regulatory compliance issues and data integration standards: CDISC, HL7, ICH

KEY QUESTIONS ANSWERED

Our research addresses the following issues that are critical to your success:

- How can both the cost and the risk of clinical trials be managed while maximizing time to market and data quality?
- What are the trends in technologies for trial design, patient recruitment, and site monitoring, and how will these trends affect end users in both medical device companies and the pharmaceutical industry?
- How can regulatory compliance be achieved while staying on time and within budget?
- What are the key software and IT infrastructure issues in clinical development?
- What are the key issues and best practices in data collection, data validation, and reporting?
- What are the best practices and tools to use to optimize trial management through accelerating

patient and investigator recruitment, managing sites efficiently, and balancing the portfolio of trials?
- What are the best practices seen from companies leading the implementation of IT in clinical trials?
- How do I ensure that my company's use of IT in clinical trials compares favourably with that of my competitors?

WHO SHOULD SUBSCRIBE

The adoption of eClinical technologies is rapidly moving from a tactical, project-based technology acquisition process to a strategic, enterprise-wide investment methodology. As a result, the insights provided from the *Clinical Development Strategy and Technology* service are beneficial to a growing number of individuals in the life science community, including chief information officers (CIOs), eClinical technology vendors, clinical operations, IT executives, and clinical investigators.

RETURN ON RESEARCH INVESTMENT

Core Benefits for Technology Buyers:
- A clear view into the benefits, as well as the risks, of adopting eClinical technologies
- The ability to select best-in-class solutions, data interchange standards, and technology vendors
- Strategies to maximize your investment in existing technologies

Core Benefits for Technology Solution Providers:
- An assessment of the market opportunity for eClinical technology solutions
- Unbiased insight from key opinion leaders in the industry
- Increased market exposure and competitive intelligence

STANDARD SUBSCRIPTION DELIVERABLES

Health Industry Insights values building client relationships and strives to foster a partnership with each client through continuous analyst interaction; concise delivery of current, timely, and relevant research; as well as custom research when needed. Each advisory service delivers strategic research covering business and technology issues. A summary of the basic service deliverables available to subscribers includes the following:

- Analyst inquiry time specific to the advisory service area
- 24 x 7 research access via healthindustry-insights.com
- Discounted admission to specific Health Industry Insights or IDC conferences
- Complimentary forum invitations

JOIN OUR CONVERSATIONS

Our Insights Communities are designed to enable technology buyers to engage with our industry analysts, share knowledge and best practices, and connect with colleagues.

To participate in our communities, members create a profile to:

- Participate in our discussions in our blogs, forums and polls.
- Learn about and share best practices, tips, tricks and tools related to business/technology.
- Network with colleagues and join groups within each industry.
- Provide feedback to IDC's industry analysts and hear what we're up to.

RELATED OFFERINGS

Related offerings to our industry-specific research practices include IT Spending Guides - segmented regional and country-level analysis by solution area; Go-To-Market Services (GMS) - a modular set of deliverables to assist clients in extending their market reach, validating their market strategy, and producing more responsive customer-centric marketing campaigns; Global IT Advisor (GITA) - a flexible, cross-disciplinary research service providing granularity in market information and analysis across a wide range of technology areas; and IT Executive Strategies (ITES) - a best-practices research service covering IT governance, portfolio management, vendor procurement, and more. In addition, Health Industry Insights experts are available for individualized consulting engagements to address specific challenges or to customise the results of our syndicated research offerings.

Four trends stand out: continued rapid growth; increased focus on innovation; differentiation of opportunities; and India's increased importance. Like bamboo, Asia's green shoots grow faster than those in the West. IMS Health predicts low single-digit pharma growth for EU and the US, but double-digit growth rates for APAC over the next five years.

With much of their incremental growth revenues coming from APAC, over the next decade, biopharma companies are shifting R&D spend into the region. This is not just to leverage the well-publicized lower costs and faster recruitment benefits, but is also about market access—as placing global trials in APAC speeds up product launches and product penetration.

For us in the CRO space, this is of course very good news. Business for CROs is predicted to grow even faster than sponsor company growth. Why? Because in addition to the shift of R&D spend to APAC, many sponsor companies simply do not have large organizations in the region. Engaging a CRO means sponsors can go straight to a more agile, and flexible outsourcing model without complications of managing legacy and headcount issues.

Growth remains the key word for APAC. Global biopharma and CRO companies are increasingly observing their success in APAC as a surrogate measure of their global success. The companies now realise that they cannot succeed globally without being successful in APAC.

Innovation is becoming a buzz word for APAC. Some of this is traditional biopharma innovation generated in the labs—and the world will see more novel compounds appear from companies such as S-Bio in Singapore. The next decade will also see APAC biopharma companies replicating the process and enabling technology innovation, which is seen in the IT and electronic industries. This innovation will act synergistically with Asian advantages for drug development.

A great example is the drug development enabling website, goBalto.com. This site (incidentally set up by a Korean entrepreneur Jae Chung) will enable Next Generation virtual drug development companies to more easily tap into the burgeoning talent in APAC. Imagine the formation of an Asia-enabled drug development machine, which leverages Asia's lower costs and faster project cycle times, to get products much faster and cheaper than in the West. This has the potential to change the global biopharma operating model in the same way that Japanese car manufacturers transformed the auto industry.

For global CROs, there's tremendous opportunity to facilitate innovation in APAC by providing a bridge between East and the West. As the region matures, greater differentiation is becoming apparent. I have always referred to the APAC region as a portfolio of countries, each with different opportunities. For example, the fast, efficient and predictable regulatory environment in Singapore allowing early site initiation, massive population in India and China providing rapid enrolment and Taiwan's pedigree in oncology studies. Now, there is more clarity about what these differences are, and experienced CROs are better able to allocate development projects across different locations to optimize the mix of benefits and minimize downsides.

Expect this trend to continue. Singapore provides a good example. Recognizing the limitations of its relatively small population, it is becoming a regional center-of-excellence in early-phase studies that require small subject numbers but a fast and predictable regulatory environment. Another 'small' populated country, which is building on an already strong reputation in the early-phase space, is Australia. Expect Australia to increase its focus on Health Economics and Outcomes Research (HEOR) work, as there is a tremendous experience base, thanks to the strict requirements of their regulatory and reimbursement environment.

CROs will also become more differentiated. People in APAC still have difficulty in explaining the differences between various CROs. However, the industry's rapid growth in the region will provide real opportunities to build differentiated CROs in the region.

CROs will be differentiated on the basis of service offering and leveraging across centers-of-excellence in different locations in APAC. For example, biometrics capability in India and China, regulatory and HEOR consulting in Australia and market access insight in China and Korea.

Differentiation also will occur with certain product related opportunities. APAC has become the center-of-excellence for oncology studies as highlighted in a recent research article. Biosimilars are an area of real opportunity for APAC with companies such as Celltrion in Korea making global impact in this growing sector.

The final trend that is increasingly prominent is India's emergence as the hot growth area in drug development. Volumes have been written and many conference-hours spent discussing India's clinical research capabilities, yet the story is still unfolding. India continues to offer new, unanticipated and interesting opportunities.

The original drivers of low cost and access to patients continue to be attractive. India is making its regulatory system more robust, stringent and effective, which will further increase the country's appeal.

Those traditional benefits will ensure India to grow rapidly in its drug development industry for many years. India is also the real source of innovation, particularly in how things can be done better and more efficiently.

One such idea is reversing the usual talent deployment approach for successful delivery of international projects. Traditional deployment of talent involved placing project leaders in more mature clinical research markets, while India/APAC's participation remained limited to lower level functions in the value chain. However, working towards maturing local talent to higher level functions in a systematic way offers several advantages including enhanced efficiency, 24/7 leadership coverage for projects and a 'can do' mindset that really resonates well with sponsors. For example, local employees at Kendle India love the career opportunities this provides, with the result, we enjoy as one of the lowest turnover rates of the employees in the industry.

Macro level trends for India includes:

- Relatively stable economy and buoyant local biopharma companies and
- Well rounded progress of clinical development functions marked by addition of newer services, and as reflected in recent strategic deals initiated between top 10 multinational companies with Indian giants.

ROLE OF FOOD AND DRUG ADMINISTRATION

As a professional couple with one half in the world of human subjects' protection and the other half in the world of academic pulmonary medicine, we read the article by Miller and Shorr and the accompanying editorial by Perkins with great interest. The article, in the April 2002 issue of *CHEST*, elucidated the many challenges faced by all of us who conduct research involving human subjects. The authors have brought into focus the ethical challenges presented to the investigator, the institutional review board (IRB), and the sponsor. Although we agree with the authors regarding the problems that they have identified, we suggest that the responsibility for the problems is not limited to the three groups they have identified. We would like to point out some of the inherent and fundamental problems of industry-sponsored randomized clinical trials as they are currently conducted.

The current practice is that industry (with some input from expert consultants) designs clinical trials, the US Food and Drug Administration (FDA) reviews them, and then the sponsor then seeks sites for these trials. As Miller and Shorr point out, clinical trials have seen a tremendous change in recent times, moving away from academic medical centers and towards private practice settings. The scrutiny of IRB applications for these trials is different given the context of an independent IRB relative to the academic IRB. There is a potential and very real problem of "IRB shopping" with the obvious benefit that sponsors can have clinical trials conducted, unopposed and without modification. IRBs, both local and independent, are under extreme pressure to review protocols quickly and without requesting modification. By contrast, there is a strong tradition and clear need for local control of IRBs so that their actions reflect the communities that they serve. However, Miller and Shorr have pointed out the universal nature of the ethical dilemma confronting the investigators, the IRB, and the sponsor. Obviously, these groups did not meet this ethical challenge in the mometasone trial.

In this context, such review standards and, in fact, this level of review would be more appropriate at the level of the FDA. At the present time, the role that the FDA plays in the risk/benefit analysis for the subject is ambiguous. This realization leads to a variety of troubling questions. What was the role of the FDA in allowing a study such as the mometasone trial to proceed? In general, what kind of review does an investigational new drug application undergo? Do experts in the area of study review it? Does the FDA consider whether the protocol meets current guidelines for the treatment of the condition under study, or does the FDA see this as the responsibility of the sponsor or the local IRB?

Our local IRB is frequently told by sponsors that we are the only IRB that has issues with a particular protocol. These issues usually relate to our concerns with including a placebo arm, or with withdrawing or withholding proven treatment. Industry-sponsored clinical trials that do not pass ethical muster or put patients at unjustifiable risk should be stopped at the FDA review. They should be rewritten according to currently accepted guidelines for the treatment of the condition under study, and according to regulations meant to protect any vulnerable subjects who might be included.

At national IRB meetings last year (Public Responsibility in Medicine and Research), Robert Temple, MD, of the FDA suggested that the agency preferred to allow protocols to have broad inclusion and exclusion criteria so as not to limit multicenter trials. He saw narrowing of inclusion and exclusion criteria to make a study safe as the responsibility of the IRB. At the same IRB meeting, a representative of the pharmaceutical industry was quick to point out that academic IRBs were a problem for industry, and they would prefer that independent IRBs review all multicentered trials. The implications of this latter statement are obvious. However, the suggestion of the FDA that narrowing criteria should be done at the local IRB feeds this somewhat dysfunctional relationship. The federal government has now begun to realise that IRB shopping is a real phenomenon.

Currently, proposed regulations to prohibit IRB shopping are available for public comment. However, prohibiting IRB shopping alone will not resolve the current problems involving the clinical trials industry that lead to conducting trials such as the one reviewed and analysed by Miller and Shorr.

Unquestionably, unethical clinical trials have been funded by sources other than industry, and have been approved by local IRBs, including academic IRBs. National Institutes of Health-funded studies have also failed the risk/benefit analysis. In this time of heightened public awareness of the shortcomings of the human subjects' protection process, we would argue that rigorous risk/benefit analysis should be applied before protocols reach the local IRBs. In this way, we can help prevent local conflicts of interest and ensure that research is conducted at a standard deserving of the public trust.

The establishment of FDA as it exists today grew out of a time early in the century when consumers were victimized by dishonest purveyors of fraudulent potions and compounds that were ineffective, dangerous, or both. A system of drug regulation was established in the United States that has served us well. Under this system, FDA reviews new drugs to assess their safety and efficacy. In addition, certain types of drugs must be prescribed and dispensed by only licensed health care professionals. The prescribing requirement is based on the principle that certain drugs have risks of such significance associated with them that they should be administered only under the supervision and recommendation of a "learned intermediary" - that is, a licensed practitioner with the education and training necessary to oversee the administration of potentially harmful drug products. Similarly, these products may only be dispensed by a licensed professional that can help to assure proper dosing and administration, and can provide important information on the drug's use to patients. The types of unlawful conduct involving online drug sales that FDA has identified are similar to unlawful activities that occur in other sales contexts. Under the Federal Food, Drug, and Cosmetic Act, FDA has the legal authority to take action against:

- The importation, sale, or distribution of an adulterated or misbranded drug;
- The importation, sale, or distribution of an unapproved new drug;
- Illegal promotion of a drug;
- The sale or dispensing of a prescription drug without a valid prescription; and,
- Counterfeit drugs.

When the Internet is used for an illegal sale, FDA, working with the Department of Justice, must establish the same elements of a case, develop the same charges, and take the same actions as it would if another medium, such as a storefront or a magazine, had been used.

FDA has investigated and referred cases for criminal prosecution and initiated civil enforcement actions against online sellers of drugs and other FDA-regulated products, particularly sellers of drugs not approved by the Agency. As will be described later, FDA has significantly expanded its enforcement activities during this past year with regard to online drug sales.

THE NEW DRUG APPLICATION (NDA)

INTRODUCTION

For decades, the regulation and control of new drugs in the United States has been based on the New Drug Application (NDA). Every new drug has been the subject of an approved NDA. The NDA application is the vehicle through which drug sponsors formally propose that the FDA approve a new pharmaceutical for sale and marketing in the U.S. The data gathered during the animal studies and human clinical trials of an Investigational New Drug (IND) become part of the NDA.

The goals of the NDA are to provide enough information to permit FDA reviewer to reach the following key decisions:

- Whether the drug is safe and effective in its proposed use(s), and whether the benefits of the drug outweigh the risks.
- Whether the drug's proposed labeling (package insert) is appropriate, and what it should contain.
- Whether the methods used in manufacturing the drug and the controls used to maintain the drug's quality are adequate to preserve the drug's identity, strength, quality, and purity.

RESOURCES FOR NDA SUBMISSIONS

The following resources have been gathered to provide you with the legal requirements of a new drug application, assistance from CDER to help you meet those requirements, and internal NDA review principles, policies and procedures.

GUIDANCE DOCUMENTS FOR NDAS

Guidance documents represent the Agency's current thinking on a particular subject. These documents are prepared for FDA review staff and applicants/sponsors to provide guidelines to the processing, content, and evaluation/approval of applications and also to the design, production, manufacturing, and testing of regulated products. They also establish policies intended to achieve consistency in the Agency's regulatory approach and establish inspection and enforcement procedures.

Because guidances are not regulations or laws, they are not enforceable, either through administrative actions or through the courts. An alternative approach may be used if such approach satisfies the requirements of the applicable statute, regulations, or both. For information on a specific guidance document, please contact the originating office. Formatting, Assembling and Submitting New Drug and

Antibiotic Applications. Submitting Supporting Documentation in Drug Applications for the Manufacture of Drug Substances.Submitting Documentation for the Stability of Human Drugs and Biologics.Submitting Samples and Analytical Data for Methods Validation.Submitting Supporting Documentation in Drug Applications for the Manufacture of Drug Products. NDAs: Impurities in Drug Substances.Format and Content of the Human Pharmacokinetics and Bioavailability Section of an Application.Format and Content of the Nonclinical Pharmacology/Toxicology Section of an Application. Providing Clinical Evidence of Effectiveness for Human Drug and Biological Products.

Describes the quantity of evidence, and the documentation of the quality of evidence necessary to support a claim of drug effectiveness.·Drug Master Files. A Drug Master File (DMF) is a submission to the FDA that may be used to provide confidential detailed information about facilities, processes, or articles used in the manufacturing, processing, packaging, and storing of one or more human drugs.·Required Specifications for FDA's IND, NDA, and ANDA Drug Master File Binders Qualifying for Pediatric Exclusivity.

Certain applications may be able to obtain an additional six months of patent exclusivity.PET Drug Applications - Content and Format for NDAs and ANDAs [HTML] or [PDF]. Refusal to File.Clarifies CDER's decisions to refuse to file an incomplete application.

LAWS, REGULATIONS, POLICIES AND PROCEDURES

The mission of FDA is to enforce laws enacted by the country Congress and regulations established by the Agency to protect the consumer's health, safety, and pocketbook. *The Federal Food, Drug, and Cosmetic Act* is the basic food and drug law of the country With numerous amendments, it is the most extensive law of its kind in the world. The law is intended to assure consumers that foods are pure and wholesome, safe to eat, and produced under sanitary conditions; that drugs and devices are safe and effective for their intended uses; that cosmetics are safe and made from appropriate ingredients; and that all labeling and packaging is truthful, informative, and not deceptive.

CODE OF FEDERAL REGULATIONS (CFR)

INTRODUCTION

Code Of Federal Regulations (CFR) The final regulations published in the Federal Register(daily published record of proposed rules, final rules, meeting notices, etc.) are collected in the CFR. The CFR is divided into 50 titles which represent broad areas subject to Federal regulations. The FDA's portion of the CFR interprets the Federal Food, Drug and Cosmetic Act and related statutes. Section 21 of the CFR contains all regulations pertaining to food and drugs. The regulations document all actions of all drug sponsors that are required under Federal law. 21CFR Part 314 - Applications for FDA Approval to Market a New Drug or an Antibiotic Drug.

CDER's Manual of Policies and Procedures (MaPPs). These documents are approved instructions for internal practices and procedures followed by CDER staff to help standardize the new drug review process and other activities. MaPPs define external activities as well. All MaPPs are available for the public to review to get a better understanding of office policies, definitions, staff responsibilities and procedures. MaPPS of particular interest to NDA applicants include:

Review of the Same Supplemental Change to More than One NDA or ANDA in More Than One Review Division. Filing Review Issues. Action Packages for NDAs and Efficacy Supplements. Refusal to Accept Application for Filing From Applicants in Arrears.Drug Application Approval 50 Policy· Requesting and Accepting Non-Archivable Electronic Records for New Drug Applications.

Despite billions of dollars, decades of research, and an unparalleled level of international cooperation between research scientists and clinicians, cancer remains a major cause of death worldwide. Therapies have improved and many forms of cancer are now treatable, but the disease still kills over six million people throughout the world each year. For this reason, cancer research funding through the National Institute of Health (NIH) and the National Cancer Institute (NCI) represents one of the largest expenditures of the United States federal government. This focus on research has led to improved medical treatments for cancer as well as greater understanding of the molecular intricacies of the disease. Cancer is now considered not simply one disease but rather a multitude of independent disorders that can all result in malignant cellular growth. Unlike other diseases, there is no simple explanation, or even definition, for cancer, and so there is no simple way to treat it. The dream of a cure for cancer, the "magic bullet" to miraculously eliminate the disease, is now considered unrealistic in light of overwhelming complexities.

However, an interdisciplinary approach has emerged offering new hope for cancer therapies. Standard chemotherapy or radiation regimens are nonspecific in their targeting of cancer cells. In contrast, this new field of therapy reflects a growing awareness on the part of the oncology community that the most effective way to treat a tumor is to target the specific molecular mechanisms responsible for its uncontrolled growth. Research has shown that not all tumors are the same; molecular targeting, the attempt to tailor therapy to the specific abnormalities causing disease, has become one of the most promising areas in cancer biology research. In addition, as a specific application of molecular targeting, pharmaceutical research has combined the basic science of molecular biology and biochemistry with combinatorial chemistry and organic synthesis to create a new field: rational drug design.

Unfortunately, drug development takes time, and it is years, often decades, before a drug that shows promise in the lab will make it into the clinic. Rational drug design, because it narrows the focus on possible therapeutic agents, is a potential way to shorten this lengthy process. The inherent benefits of the approach are highlighted by one of the most exciting developments in cancer biology in recent years. On May 10, 2001, as a result of overwhelmingly positive clinical trials, the Food and Drug Administration (FDA) approved Novartis Pharmaceutical's Gleevec (imatinib mesylate) for use in the treatment of chronic myelogenous leukemia (CML). In a stunning validation of the value of rational drug design, Gleevec is the first of a wave of compounds with the potential to revolutionize the way we look at cancer.

THE TARGET - CHRONIC MYELOGENOUS LEUKEMIA

Leukemia is a disease of the white blood cells. While normal white blood cells, or leukocytes, are responsible for maintaining the immune system, the abnormal proliferation of such cells leads to a deadly form of cancer. Chronic myelogenous leukemia (CML) is characterized by the accumulation of immature granulocytes, a specific type of white blood cell. The disease progresses through three stages, and the specific stage of disease in a patient is crucial in planning treatment. In stage one, also known as chronic phase, there are few if any outward symptoms of disease; many CML patients are unexpectedly diagnosed as the result of blood work following a routine doctor's appointment.

After a period of up to several years, chronic phase CML progresses to accelerated phase CML, which is characterized by an increasing ratio of malignant to normal white blood cells. The final stage of CML is known as blast crisis, and by this point almost a third of blood and bone marrow cells are cancerous. Due to their high numbers, some of these blast cells may metastasize and form tumors in other tissues including the bones or lymph nodes. Until recently, the most effective form of treatment for CML was the use of a compound called alpha interferon which was often followed by high dose radiation, chemotherapy, and a bone marrow transplant. However, as CML progresses, the efficacy of bone marrow transplants decreases sharply. Transplants done in the accelerated or blast crisis phase have a high rate of failure. Only 15% of patients transplanted in blast crisis are still living after five years (Leukemia Society). Despite the grim prognosis for CML patients, there is one crucial molecular detail of the disease that renders it a promising target for rational drug design. The highly specific targeting that led to the laboratory development of Gleevec can be traced back to some of the early studies of leukemia. The first piece of the puzzle was uncovered in 1960 when it was discovered that nearly all patients with CML possess an abnormally small chromosome 22. This abnormality was nicknamed the Philadelphia chromosome and has since become known as CML's cytogenetic calling card.

Shortly thereafter it was discovered that chromosome 9 was elongated by approximately the same length as the missing region of chromosome 22; these findings were eventually linked after molecular characterization confirmed a translocation between chromosomes 9 and 22. In the process of this exchange, breakpoint cluster region (bcr) proteins from chromosome 22 become fused with the c-abl oncogene from chromosome 9 thus creating a hybrid bcr-abl gene. A normal c-abl gene encodes a protein that is highly regulated; the functional enzyme is shuttled between the nucleus and cytoplasm of a cell so that it is only activated when needed. However, under the influence of bcr resulting from bcr-abl fusion, c-abl loses the checkpoints intended to hold it under control. The enzyme becomes a constitutively active cytoplasmic protein. Studies in mice have shown that the introduction of the bcr-abl fusion protein, regardless of the presence of the Philadelphia chromosome, is sufficient to experimentally induce the development of CML.

The consequences of the bcr-abl abnormality are devastating because of the resulting c-abl enzyme that is produced. This unregulated enzyme, a tyrosine kinase, is the molecular cause of CML. Kinases are enzymes that transfer phosphate groups from ATP to particular residues, in this case tyrosines, of target proteins within the cell. Phosphorylation often serves as a molecular switch to activate or inactivate a range of proteins involved in such vital activities as cellular growth, development, and survival. For this reason, kinases are normally kept under tight control; a runaway kinase can promote the excessive proliferation associated with cancers including CML.

THE GLEEVEC APPLICATION

Since abnormal enzymes are implicated in a variety of human cancers, targeting the mutant function should have deleterious affects for the tumor while leaving normal tissue unaffected. As a result of studies highlighting the molecular mechanisms of CML, a search began for compounds to inhibit the runaway *bcr-abl* kinase. Compound screening at Novartis Pharmaceuticals in the early 1990's identified 2-phenylamino-pyrimidines as a general class of kinase inhibitors, and subsequent chemical synthesis was used to design a series of compounds that should have more specific anti-tumor effects. One of the drugs in this class was called STI571.

In vitro studies using STI571 demonstrated that it was a highly effective and specific inhibitor of platelet-derived-growth-factor (PDGF) kinases, c-kit kinases, and c-abl kinases. STI571 acts by blocking the ATP-binding site of these enzymes. The functional form of the enzyme uses ATP as its source of phosphate groups, thus by blocking the ATP-binding site of the kinase, STI571 prevents the enzyme from becoming active. In the case of bcr-abl kinases, further experiments demonstrated that STI571 inhibited growth in cell lines carrying the bcr-abl mutation. Animal models produced data indicating an STI571 dose-dependent inhibition of tumor growth although tumors were not completely eliminated.

On the basis of promising laboratory data, STI571 began clinical trials in 1998. The first trials were conducted in patients with chronic phase CML who had previously failed interferon therapy. As with any Phase I clinical trial, the main purpose of the study was to determine the overall safety and acceptable dose ranges of the drug. However, in the midst of safety evaluations, the physicians running the trial noticed that with elevated doses of STI571 an astounding 98% of the patients showed a complete hematologic response - the abnormal Philadelphia chromosome was still detectable, but the elevated levels of circulating white blood cells, the cause of CML symptoms, were gone. Due to these astonishing results, more expansive Phase II trials were conducted in over 1000 patients representing a range of CML stages, 27 cancer centers, and six countries. In the 532 chronic phase patients, the results of the original trial were confirmed. Only 3% of patients experienced disease progression while a mere 2% did not complete the trial due to drug side effects. In addition, of the 86% of patients on long term STI571 treatment, 75% achieved either complete or major cytogenetic responses (Mauro and Druker). Not only did blood counts return to normal, but in many cases the molecular causative agent of disease, the Philadelphia chromosome, had become undetectable.

These results were also observed in the 233 accelerated phase patients: 91% achieved a hematologic response, 41% had a cytogenetic response, and mean survival at one year was 74% (Mauro and Druker). In blast crisis patients, typically the most difficult group to treat, 64% of the 260 patients achieved either a hematologic or cytogenetic response while the median survival of 6.8 months compared favourably with the 3 month median survival for bone marrow transplant recipients (Mauro and Druker). As a result of its clear clinical benefits for the treatment of CML, in May, after a mere three months of review, the FDA approved STI571 for general use under the trade name of Gleevec. The speed of this decision reflects the amazing significance of Gleevec's application for the treatment of an often fatal disease.

LOOKING TO THE FUTURE

As a chemical agent, Gleevec clearly represents an important advance in the field of cancer therapy. It has minimal side effects, mostly limited to mild nausea and edema without involving the major systemic toxicity and organ-targeting common to many chemotherapy agents. The drug can also be taken orally, which not only frees the patient from the painful process of repeated intravenous administration of chemicals but also eliminates the requirement for hospital admission. In research applications, Gleevec's specific targeting and almost unbelievable elimination of cancerous cells represents solid support for rational drug design programmes. In addition, it provides a much-needed infusion of hope for a field that can often seem hopeless, particularly in light of the dreams of a "cure for cancer" which have yet to be truly realised.

Despite Gleevec's enormous significance and established clinical benefits, it is still not a complete panacea. There have been reports of CML cases with established Gleevec resistance (Marx). As with other forms of therapy-resistant cancers, additional research is necessary to maximize the efficacy of CML treatment. In one possible improvement, Gleevec is being combined with existing therapies in an attempt to circumvent the drug resistance problem (Marx). Further studies are needed to determine Gleevec's long term health effects on patients as well as its effects on the cancer itself. The drug is still so new that it is uncertain how long the remarkable therapeutic effects will persist. In many CML patients, Gleevec treatment results in an apparent partial or complete elimination of the abnormal Philadelphia chromosome.

It is unknown whether cessation of drug treatment will result in a recurrence of the bcr-abl fusion. In addition, while Gleevec is a spectacular example of the power of molecular targeting, it does have a distinct advantage in that CML is almost unilaterally caused by a single molecular event. Most other cancers are far more complex. The true utility of molecular targeting will be tested in the development of therapies to treat cancers resulting from multiple molecular aberrations.

Yet even in this arena Gleevec may still prove useful. While the complete ramifications from clinical studies are pending, Gleevec's kinase inhibitory abilities have been tested against the c-kit and PDGF kinases. C-kit and PDGF mutations are known to be a part of the progression of certain kinds of gastrointestinal, prostate, and lung cancers as well as glioblastoma, an aggressive form of brain cancer (Mauro and Druker). Cancer's molecular complexities and devastating consequences ensure that oncology research will remain a challenging and vitally important intersection of medicine and basic science. Researchers and clinicians must continue to effectively utilize new laboratory techniques, experimental knowledge, and clinical evidence in the advancement of cancer therapy. Gleevec's fundamental contribution to this process is the validation of rational drug design. As illustrated so clearly in the treatment of CML, molecular targeting can be used to effectively treat cancer. In the words of Dr. Richard Klausner, the director of the National Cancer Institute: "Gleevec offers proof that molecular targeting works in treating cancer, provided that the target is correctly chosen. The challenge now is we've got to find these targets." Gleevec's legacy is most clearly seen in a new direction of cancer research: the quest for molecular targets.

5

Vitamins

WATER AS SOLUBLE VITAMINS

This manuscript describes the lecture on vitamins contained in the core course entitled,"The Biochemical Basis of Drugs and Diseases. In the first year, our curriculum is designed to focus on systems and diseases. As such, this course acts in concert with both the Anatomy and Physiology course and the Pathophysiology course to present an integrated view of human diseases and systems. The second year of our curriculum is focused on learning about drugs classes and the mechanism of action.

The lecture on vitamins is the last topic of a course focused primarily on metabolism. Because many vitamins play major roles in metabolic cycles, this lecture allows for a brief review of much of the material covered throughout the course. Therefore, vitamin examples are primarily chosen to reinforce major metabolic cycles and also their role in human disease. There is more clinical information that is relevant to vitamins than what is presented in this lecture. However, since first year students in our curriculum have almost no knowledge of therapeutics, the focus of the lecture is on diseases and not clinical practice.

INTRODUCTION

Vitamins are a multibillion dollar industry.1 They are readily available to the public and are the focus of the most frequently asked questions to pharmacists.2 Surprisingly, most pharmacy students receive little training on the many roles of vitamins in nutrition.2 Therefore, this lecture attempts to not only reinforce fundamental biochemical and metabolic pathways in the human body, but also to provide pharmacy students with practical information. The diet is a vast source of important nutrients.

These nutrients include several important classes of biomolecules such as:

- Energy yielding components (carbohydrates, lipids, and proteins).
- Essential and nonessential amino acids.
- Essential fatty acids.
- Minerals.
- Vitamins.

Vitamins are organic substances that must be provided by the diet either because they cannot be biosynthesized or the amount that is provided through biosynthesis is inadequate for maintaining normal health. Vitamins are broadly divided into 2 classes based upon their hydrophobicity. The more hydrophilic vitamins are termed the water-soluble vitamins and are composed of the B-complex vitamins and vitamin C. The more hydrophobic vitamins, referred to as the fat-soluble vitamins, are composed of Vitamins A, D, E, and K. This article focuses primarily on the watersoluble vitamins due to their greater role in the major metabolic cycles, which are the primary focus of this course.

The normal North American diet is sufficient to prevent significant vitamin deficiencies and the related diseases associated with these deficiencies. However, there is increasing concern that slight vitamin deficiencies in a number of water-soluble vitamins (B1, B6, B12, folate and vitamin C) are risk factors for diseases such as such as depression, cancer, cardiovascular disease, and neural tube defects. The prevalence of this slight vitamin deficiency is likely related to diet, since most of these vitamins are supplied by fruit and vegetable intake. A recent survey has shown that only 25% of the population meets their daily intake of 5 servings.3 Food preparation is another source of vitamin loss. For example, heating food for more than 2 hours causes more than a 10%

loss of most water-soluble vitamins. Refrigeration, freezing, and reheating have all been shown to lead to further loss of vitamins. Exposure to light causes significant loss of riboflavin from foods and the combination of heating and light can almost completely remove this vitamin from food.

Thiamin (Vitamin B1)

Thiamin functions in the body as thiamin pyrophosphate (TPP) an important enzyme co-factor.5 Thiamin reacts with adenosine triphosphate (ATP) to form thiamin pyrophosphate through a reaction mediated by the enzyme thiamin diphosphokinase. Following its production, TPP is incorporated into 2 important enzymes: pyruvate dehydrogenase and ketoglutarate dehydrogenase.

Pyruvate dehydrogenase is part of a multi-enzyme complex that acts to convert pyruvate generated in glycolysis into acetyl-CoA for entry into the tricarboxylic acid (TCA) cycle. Thiamin pyrophosphate is also used as a co-factor for the enzyme -ketoglutarate dehydrogenase, which is a key point of regulation in the TCA cycle. -Ketoglutarate dehydrogenase is involved in the conversion of ketoglutarate to succinyl CoA. In the case of both enzymes, TPP assists in decarboxylation of a small ketoacid.

In addition to its uses in metabolism, thiamin may enhance circulation and blood formation. It is required for maintenance of the nervous system and is used in the biosynthesis of the neurotransmitters acetylcholinc and *hydroxybutyrate (GABA). Thiamin also is used in the production of hydrochloric acid in the stomach and, therefore, has a role in digestion.

Thiamin is provided in the diet through most grains. A deficiency of thiamin is called beriberi and it is most often observed in Southeast Asia.6 The main staple of diet in this part of the world is rice, which does not provide dietary thiamin to as great an extent as other grains. The symptoms of beriberi include limb pain, muscle weakness, and low cardiac output. all symptoms are related to the diminished capacity of the major energy producing pathways that are dependent on TPP as a co-factor. Mild deficiencies of thiamin are sometimes observed in the elderly and in low-ıncome groups on restricted diets. The earliest symptoms of thiamin deficiency are loss of appetite, constipation, and nausea. Other symptoms such as mental depression, peripheral neuropathy, irritability, and fatigue are related to the role of thiamin in maintaining a healthy nervous system.

The most common form of thiamin deficiency in the United States is alcoholic neuritis. Because alcoholics normally have a poor appetite, overall food consumption is low. In addition, alcoholics are predisposed to developing nutritional deficiencies since alcohol is their primary calorie source. Under the broad umbrella of alcoholic neuritis come two disorders, Wernicke's Syndrome and Korsakoff s Psychosis.7 The symptoms of Wernicke's Syndrome (ophthalmoplegia, nystagmus, and ataxia) respond quickly to administration of thiamin, whereas the more severe memory and learning disorders associated with Korsakoff s Psychosis are refractory to thiamin treatment.

Riboflavin (Vitamin B2)

The reduced forms of these enzyme co-factors can donate these electrons to return to their previous fully oxidized forms. It is this ability to act as a conduit for electron transfer reactions that makes FAD and FMN such important enzyme cofactors.

Riboflavin is available from a wide variety of dietary sources such as milk, cheese, meat, eggs, and cereal products. Symptoms associated with riboflavin deficiency include sore throat, dermatitis, anemia, neuropathy, and cataract formation. Riboflavin deficiency is not normally observed in the United States; however, deficiency may be observed as part of a general case of malnutrition or in cases of chronic alcoholism.8 The conversion of riboflavin to FMN is required for absorption and transport into many tissues.

The conversion has been shown to be inhibited by hypothyroidism and the structurally similar medications chlorpromazine, adriamycin, quinacrine, tetracycline, and tricyclic antidepressants.9-12

Niacin (Vitamin B3)

Niacin can be found in foods such as meats, breads, and beans. Mild niacin deficiencies have similar symptoms to those observed with riboflavin, which is not surprising due to the similar roles both of these vitamins play in biochemical reactions. Niacin deficiencies are occasionally observed in alcoholics, cases of general malnutrition, and in the elderly on restricted diets.

A severe deficiency of niacin is known as pellagra, which is derived from the Italian phrase meaning rough skin. Pellagra is marked by dermatitis and also is notable for causing a blackening of the tongue. Early cases of pellagra were first observed in Europe shortly after the introduction of corn from the voyages of Christopher Columbus. In these cases, poor farmers who were raising corn as animal feed were particularly susceptible. Because niacin in corn is not bioavailable unless treated with a strong base such as lye, the farmers developed niacin deficiencies.

Europeans did not know this processing method until revealed to them by Native-Americans during the early colonial period of North America. However, the connection between pellagra and niacin was not known until the early 20th century. Pellagra was a significant health issue in the United States over the period from 1900-1940 resulting in over 100,000 deaths. Today most diets are supplemented with niacin through enriched flour, which receives its name because of the added niacm.13

Niacin can be administered in doses of 2g to 4g to causes a decrease in circulating levels of cholesterol and LDL. While the cholesterol lowering effects of niacin are desirable there are potential side effects from such large doses of this vitamin. The most immediate reaction observed from large doses of niacin is vasodilation resulting in flushing. Over time there may be a reduction in fatty acid mobilization causing a depletion of glycogen and lipid stores in muscle tissue. Long-term exposure may also elevate blood glucose and uric acid levels, suggesting increased risk for patients who are on the borderline for diseases such as diabetes and gout. Prolonged use of high doses of niacin can lead to elevated levels of the serum enzymes alanine aminotransferase and aspartate aminotransferase, which may suggest liver damage.

Pyridoxine (Vitamin B6)

Pyridoxine is the precursor to the active enzyme cofactor pyridoxal phosphate (PLP). Pyridoxal phosphate is a critical co-factor for enzymes involved in reactions involving many amino acids. The N-terminus of the amino acid forms a covalcnt bond to PLP, allowing a wide variety of displacement reactions to occur at the alpha carbon. These include decarboxylations, transaminations, and transfers of side chains. Pyridoxine, therefore, plays a central role in the production of many neurotransmitters, such as serotonin, norepinephrine, and histamine. Pyridoxine also is important in the production of heme.

The enzyme LAAAD is present in both the gut and in the brain. This presents a problem because L-dopa is readily transported across the blood brain barrier while dopamine is not. Therefore, orally administered L-dopa is rendered ineffective in the gut by LAAAD. Co-administration of carbidopa resolves this dilemma by inactivating LAAAD in the gut. Like dopamine, carbidopa cannot cross the blood brain barrier leaving LAAAD in the brain free to carry out the production of dopamine at the site of action.

In addition to hydrazine containing drugs, penicillamine, used in the treatment of Wilson's disease, cystinuria, and rheumatoid arthritis reacts with and inactivates pyridoxal phosphate.15 Patients treated with penicillamine occasionally develop convulsions, which can be prevented by supplementation with vitamin B6.

Pyridoxine is found in foods such as meats, breads, eggs, soybeans, and many vegetables. A deficiency in pyridoxine can cause facial lesions, depression, peripheral neuropathy, and glossitis.

The neurological complications can be directly linked to the effects on neurotransmitter production. Mild pyridoxine deficiencies are sometimes observed in young women taking oral contraceptives. Deficiencies have also been observed in patients with gasteroenteritis or Crohn's Disease, presumably due to poor absorption. Mild deficiencies are of concern due to a correlation with an increased incidence of breast cancer and to the risk of coronary heart disease.16,17 Neurotoxicity has been noted with doses of vitamin B6 in excess of 500 mg/ day.

Cobalamin (Vitamin B12)

Cobalamin is vital for cell growth and replication. Its major site of action is at the interface between the folic acid cycle and the active methyl cycle, where cobalamin is a co-factor for the enzyme homocysteine methyltransferase. This enzyme catalyzes the transfer of a methyl group from tetrahydrofolate to homocysteine, forming the amino acid methionine. These 2 cycles impact many other pathways due to the large number of memylation reactions in the body, especially nucleic acid biosynthesis and neurotransmitter biosynthesis.

Cobalamin is found in meats and dairy products. Deficiencies are not common because the liver can store a 6-year supply of cobalamin. In addition, cobalamin is highly conserved by enterohepatic recirculation. Strict vegetarians (vegans) may take up to 20 to 30 years to develop a deficiency, whereas inadequate absorption from the ileum due to ileitis or loss of a glycoprotein that complexes with cobalamin prior to absorption may take from 2 to 10 years to become symptomatic.

The anesthetic nitrous oxide inactivates cobalamin and can cause patients with marginal serum levels to develop deficiencies within a week.19 While rare, deficiencies in vitamin B12 are severe and manifest in erythrocytes and nervous tissue. In erythrocytes, nucleic acid biosynthesis is slowed. This results in stem cells that are committed to development into erythrocytes possessing twice their normal protein content but lacking the necessary nuclear material for proper cell division. The resulting anemia, termed megaloblastic anemia, can be quite severe.

Cobalamin deficiency can also cause permanent damage to the nervous system. The results are swelling of neurons and demylination of nerve cells, followed by cell death. These progressive manifestations cause a wide range of neurological symptoms including unsteadiness, decreased reflexes, paresthesias of the extremities, and ultimately confusion, memory loss, hallucinations, and psychosis. The neurological symptoms of cobalamin deficiency may be mistaken for multiple sclerosis. In the elderly and in alcoholics, cobalamin deficiency should be considered as a possible cause of dementia even in the absence of anemia. The neurological symptoms associated with cobalamin deficiency are not directly related to irregularities in the formation of erythrocytes.20

Nutrient depletion or shortage may arise from inadequate food intake, impaired digestion, inefficient absorption or malabsorption, inefficient assimilation and other physiologic or functional disorders. Genetic errors may interfere with or prohibit assimilation and utilization. Gastrointestinal motility disorders may have similar effects. Stress may affect both gastrointestinal tract transit times and quantities or ratios of nutrient needs, leading to depleted nutrients. Americans now spend $2.5 billion yearly on vitamins and minerals, up from $328 million 25 years ago. Roughly 55% of women and 45% of men take some kind of vitamin supplement.

Most of the vitamin supplements consumed by Americans, including those used in"enriched" foods, are making them more deficient! The greatest and most enduring of medical discoveries of the twentieth century were the vitamins. Here, at last, was the discovery of factors that caused health. Vitamin A (1912, the first vitamin to be discovered) was named retinol because, without it, a healthy retina in the eye could not be formed. Then it was found to cure night blindness. And that was just the beginning.

Along came the B-vitamins and the cures for beriberi, pellagra, pernicious anemia, nerve degeneration, enlarged heart, energy production, and much more. These came from unprocessed, whole, natural foods. When man Realised his foods had been devitalized from processing and refinement, he decided to replace the lost vitamins. But, in his effort to extricate himself from the refined food trap,"scientific confusion" led him to try to replace them with"chemical, man-made vitamins." So, again, he refined down natural vitamins into pure, crystalline vitamins, thus robbing them of important food values, just as he had refined natural grains and sweets into their pure, crystalline, carbon-white flour and sugar.

Then, to save money, he duplicated these crystalline vitamins synthetically, mostly from COAL-TAR products. Thus, natural vitamins as well as natural foods were robbed of nature's life and valuable elusive nutrients.

The latest round in conventional medicine's ongoing attempts to discredit (and ultimately outlaw) nutritional supplements is found in a highly questionable study published which claims that vitamins actually increase the risk of death.

The study claims to have Analysed a collection of previous studies on Vitamin A, beta carotene, Vitamin C, Vitamin E and selenium, concluding that most of the nutrients are actually dangerous to human health. Of course, this is research from conventional medicine - an industry that promotes patented chemicals as perfectly safe, even though FDA-approved pharmaceuticals are killing at least 100,000 Americans each year. (Imagine the uproar if vitamins killed even a fraction of that number)

To avoid getting hoodwinked by questionable research on"vitamins," you have to strongly consider the financial interests of the source of this research. JAMA accepts millions of dollars in advertising from drug companies each year, and its pages are absolutely packed with drug ads. The American Medical Association, for its part, has long worked to discredit alternative medicine and has even been found guilty by U.S. federal courts of engaging in a conspiracy to destroy chiropractic medicine. The AMA, which is largely considered a joke by anyone familiar with natural health, is hardly a credible source for publishing scientific findings on nutrition. To protect the multi-billion dollar drug industry, it's likely the AMA would say practically anything. This research published in JAMA does remind us of one important point, however:

synthetic chemicals are harmful to human health. If you take cheap"vitamins" made of these synthetic chemicals, you are doing yourself more harm than good. These cheap vitamin manufacturers, by the way, are usually owned by pharmaceutical firms.

Conventional medicine researchers try to blur the line between"junk vitamins" and"quality vitamins" by classifying ALL nutritional supplements as"vitamins," regardless of what they're really made from. By discrediting a few synthetic chemicals, they can effectively dissuade the masses from taking ANY vitamins, including the good ones. And that is, of course, their goal: to use F.U.D. (fear, uncertainty and doubt) to scare consumers away from nutritional supplements so that patients will flock to the patented, synthetic chemicals that earn drug companies billions of dollars in profits. Drugs make money for Big Pharma, and vitamins compete with drug sales.

Round up 100 people who are taking multiple pharmaceuticals, and compare their health to 100 people who are taking vitamins and nutritional supplements. Guess who's healthier? The supplement crowd will be healthier every time. It's the obvious question: If vitamins are so dangerous, where are all the dead vitamin takers? And if pharmaceuticals are so safe, where are all the super-healthy prescription drug patients? They are nowhere to be found.

The healthiest people, by far, are those who take supplements, who engage in regular exercise, and who avoid taking prescription drugs. Only fools believe research about nutrition that comes from the American Medical Association or its journal. Conventional medical researchers declaring that vitamins are worthless is about as credible as Bush Administration climatologists claiming there's no such thing as global warming. With the publication of this research, the distortion of health reality is now complete. According to the Medical Association, vitamins will kill you but pharmaceuticals will make you healthy.

DESCRIPTION ABOUT VITAMINS

It is a complex of biological"wheels within wheels"

consisting of enzymes, coenzymes, antioxidants and trace mineral activators. Enzymes, being protein, contain amino acids and trace minerals. Every mineral and amino acid needed by living cells may be found in a natural assemblage of vitamin concentrates obtained from wholesome foods. Identifying a vitamin in terms of a single chemical structure is self-defeating--you need the whole complex to get the vitamin function.

The natural vitamin complex carries trace mineral activators and without them, the vitamin fails in its metabolic function. Copper in its organic, biologically active form tyrosinase, is the trace element activator of the vitamin C-complex. Likewise, manganese activates the B complex. In turn, selenium is the trace mineral activator of the vitamin E-complex. Enzyme activators, destroyed by heat, steam sterilization and pasteurization, include Manganese, Cobalt, Molybdenum, Zinc, Selenium and Vanadium...and are effective only in the organic state.

By reducing a vitamin complex to a single chemical, chemists can claim to duplicate it in a Labouratory and easily mass produce it and sell it as an additive to food or a pill. This wouldn't be possible if they were charged with keeping all the synergistic components intact. As such, vitamins are given single chemical names and single chemical structures for the purpose of selling synthetic vitamins.

The failure of the pharmaceutical approach to vitamins originated in the chemists' synthetic isolation of aelements, uncombined with its biological matrix. The discovery of the vitamins coincided with the boom in industry, agriculture, and chemistry, which were crowding their way into all facets of life. But in the attempted duplication of living things, like artificial fertilizers or vitamins, chemistry had its most influence. The biological activity of synthetic and natural vitamins is not identical. Anyone who has studied anatomy, physiology, neurology and biochemistry will readily admit that the human body is a neuro-biochemical plant--a composite of tissues, fluids and organs that exist in a living state by virtue of neuroelectric and biochemical reactions. Being biochemical in nature, the body operates on biochemicals--LIVE

CHEMICALS--in contrast to dead or lifeless chemicals. Those biochemical factors are derived from whole, natural, uncooked and unprocessed foods. Cooking and processing alters biochemicals to dead chemicals. The human body cannot properly function if it is fueled with dead chemicals rather than live, complex biochemicals.

The difference between the live biochemical food factor and the dead chemical (counterfeit food) is dependent upon two primary requisites:

- Unaltered proteins.
- Viable enzyme systems.

Dr. Casmir Funk, who discovered the method for concentrating vitamin B, and the one who coined the term, Vitamine, said,"The synthetic product is less effective and more toxic." Every process used to extract vitamins from foods, destroys the vitamin complex. What you end up with is similar to a banana peel without the banana. When you take these incomplete supplements, they rob from the body's nutritional reserve what is missing (the banana), to try to complete the complex.

This creates what is called a rebound deficiency. If you find the shed skin of a snake, you haven't found a snake. In theBiomedical Nitty Gritty, Vol. 3, No. 1, Jan. 1982, Dr. Richard Murray D.C., presented an abundance of incontestable evidence that there is a tremendous difference between natural and synthetic vitamins..."Synthetic vitamins are reverses conformationally or mirror images of natural vitamins.

However, synthetic vitamins are not vitamins at all but synthesized fractions of a vitamin complex...a mirror image duplicate of just a portion of the real, biologically active and physiologically-precise, nutritional complex. There is no way that a fraction of a vitamin can be called a vitamin." He continues,"Textbooks record that vitamin B1 was first isolated in the year 1926. About 10 years later, scientists were able to synthesize the mirror image of Thiamine from coal-tar.

About the beginning of WWII enrichment of refined, devitalized flour became compulsory. Bleached wheat flour has more than 30 known nutrients removed with 4 synthetics added back in. Today, the majority of people in the civilized world are becoming victims of degenerative disease resulting from chronic poisoning and malnutrition. Even though well fed, they are starving to death for the most essential micronutrients. Since World War II, the American people and people of other countries as well, have had a daily ration of a genetic poison in most of the bread, flour products, cereals and other food items that are forced, by law, to be enriched with synthetic vitamins.

Then, after there develops a suspicion that these'foods' are inadequate to support life, modern advertising science steps in to propagandize the people into believing that there is nothing wrong with counterfeit or devitalized foods, that they are products of value...and...the confused public is totally unable to arrive at any conclusion of fact. In turn, they continue to buy the rubbish that is killing them off, years prematurely."

Gilbert Levin discuss optical isomers or asymmetrical molecules with atomic arrangements that are mirror images of each other, states:"Because its structure is reversed, a left-handed molecule can't take part in chemical reactions meant for a right-handed molecule any more than a left hand can fit in a right-handed glove--its odd geometry prevents it from being metabolized by the body."

In turn, the human physiology cannot properly utilize synthetic (mirror-image) fractions in the way natural complexes find their way into the biological reactions that are essential to tissue repair and the sustenance of life. A synthetic vitamin fraction can only be utilized for a drug or pharmacologic effect. Drugs are poisons, by definition. All pharmacologic therapies induce side effects and high-dose nutritional therapies are no exception. Only synthetic supplements achieve drug status.

The effect of a drug is palliative not curative. Palliation means a masking or covering-over of symptoms, while the disease process remains unchanged or progressively gets worse for lack of attention."It is better to have no vitamins at all than a mixture of crystalline-pure vitamin fractions, with one or more members of the complex missing."

Natural complexes differ from synthetic or crystalline vitamins in many ways:

- Natural complexes are colloidal, protein in nature and usually exist as enzymes or coenzymes.
- The crystalline vitamin can't be separated from its protein part without destroying its biological activity.
- The natural"live" complex carries trace mineral activators...without which the enzyme fails as a biochemical catalyst.
- Crystalline vitamins, taken into the body, must recombine (if possible) with their natural required components (robbed from the body's reserve) before they can function as a vitamin. Most, if not all of the crystalline component is lost through the kidneys.

Vitamins are not meant to be eliminated in urine any more than silverware is to be thrown out with the leftovers. Since most poisons stimulate circulation in the human body for the purpose of moving blood through the kidneys, more often to filter out the said toxin as quickly a possible...the speeded-up circulation also delivers more blood and more oxygen to the brain with a created sense of euphoria. Natural and low-potency vitamin complexes do not and will not produce euphoria, any more than eating a (non-poison-sprayed) carrot.

A synthetically-derived substance may cause a reaction in a chemically susceptible person when the same material of natural origin is tolerated, despite the two substances having identical chemical structures. The layman who doesn't understand the biochemical nature of true nutritional changes, is easily deceived by the pseudo-nutritional therapist and believes the initial response (drug effect), to the massive dose of one or more prescribed megavitamins, represents that what is being taken is exactly what is needed.

Nutritional therapists know after a time on synthetic vitamin fractions, the biochemical reaction is reversed, the euphoria ends and the original symptoms return; except amplified by vitamin poisoning. Also, the diagnosis and evaluation of symptoms are now clouded by the weeks, months or years of improper supplementation. Why the great mystery about all of this? Why don't doctors know about vitamins or nutrition in general? One big answer was revealed in the Journal of the Medical Association. It honestly admitted:"Nutrition has been neglected by the medical profession." Most medical schools devote less than three hours of total instruction in nutritional deficiency and therapy.

In short, Physicians in the United States are not required to understand nutrition to be licensed to practice medicine. It's not hard to understand why, when medical experiments are performed with synthetic vitamin fractions, these counterfeits fail to produce impressive results. But a reductionist seeks to understand the whole by dismantling its parts.

This precludes the wholistic principle of synergism, that the whole is greater than the sum of its parts. For example, what if we tried to understand the properties of water (H2O) by testing hydrogen separately and oxygen separately. Presumably, we would learn enough about each to understand what their combined characteristics would be. Combined as 2 parts hydrogen and 1 part oxygen, H2O will extinguish fire. Separately, the elements are among the most flammable and explosive elements in the universe.

The functions are exactly the reverse in their isolated state from their organically combined state. Similarly,vitamins function as biological mechanisms only when whole and complete, combined with their synergists, as in whole food. Isolated into synthetic chemicals, they fail as catalysts.

CATEGORIES OF SUPPLEMENTS

The following is a summarization of the basic differences between the four categories of supplements available today.

Pure, Hypoallergenic or Natural

This is the most prevalent category of professional supplements available today. Contrary to popular belief, the vitamins and minerals used to create supplements in this category are not derived from consumable food or botanical sources. Instead, they are synthetically produced, fractionated or isolated from refined raw materials such as crushed rock, petroleum by-products and organic solvents to produce a pure, crystalline vitamin or mineral analogue. Although the final compound is an analogous structure to a specific vitamin or mineral, it is no longer attached or associated with any of the synergistic co-factors that the nutrient is inherently found within foods. As a result, pure vitamins and minerals have significantly reduced biological activity and a lower rate of utilization in the body.

Food-Based

This category of supplements is created from a physical mix of pure, fractionated nutrients with varying amounts of food powders or extracts. There are several different factors that need to be considered when determining the absolute nutritional value of a food-based formula. The primary factors that affect the quality of a food-based supplement are the amount, variance and form of the foods included in a particular formula.

A premium food-base formula is one that provides a broad range of unadulterated whole food or plant concentrates (preferred over powders) in at least a 3:1 ratio of food to nutrients in a formula will significantly improve the bioavailability and biological activity of the vitamins and minerals. Delivering isolated vitamins and minerals in a premium food base is an energy-saving process, as the body will not waste vital energy stores to transform the vitamins and minerals into absorbable nutrients.

Ascorbic acid is a good example of a chemically synthesized vitamin that has been vastly studied; it has demonstrated enhanced utilization when the nutrient is delivered with its natural co-factors, such as bioflavonoids, rutin and tyrosinase, which always co-exist with vitamin C in food. The utilization factor can be increased by as much as 3 to 5 times (from 10% for vitamins in the first category, to 3 – 50% for food-based).

To determine if a supplement has a food-base, review the label for chemical names, as well as a combination of food extracts or powders. To determine the ratio of nutrients to foods, add up the total milligrams of the listed nutrients and for the food extracts.

Biocultured Supplements

Biocultured supplements are a physical mix of food, pure isolated nutrients and may have one or more strains of yeast and/or probiotics. Once combined, this mixture is allowed to ferment in a controlled environment. The process of fermentation, also known as culturing, biologically alters the biochemistry and nutrient content of the food. Cultured foods have been shown to have numerous health benefits and have played a role in the diet of many cultures for centuries; however, the body cannot sustain optimal health on cultured foods alone. This is due to the loss of substances, nutrients or compounds that are protective of health, such as vitamin c. Depending on the strains used during the culturing process, the generation or inactivation of toxic substances may also occur. In addition, many proteins and enzymes are denatured during the culturing of biological activity of the nutrients.

Supplements that fall into this category will denote the foods on which the yeast and/or probiotics feed. Not all cultured supplements list the isolated vitamins and minerals used in a particular formula.

WHOLE-FOOD SUPPLEMENTS

Many companies tout that their supplements are made from whole-foods; however, there are only a few companies that actually can substantiate this claim. Authentic whole food supplements are made with vitamins and minerals delivered in whole-food concentrates. The food concentrates selected have been grown in a controlled environment where budding plants are fed a nourishing broth enriched with bioactive peptide carriers. The peptide carriers act as chaperones, which govern a vitamin's or mineral's capacity to store the nutrient within its cells is successfully achieved, the growing process is complete, and harvesting is commenced using proteolytic enzymes to break down the cells' walls, and release the intracellular material. The whole-food extract or concentrate produced is rich in a particular vitamin or mineral and is delivered in a synergistic state, with all of the plant's inherent food attachments intact and unadulterated.

These attachments are vital co-factors facilitating the body's natural ability to recognize and utilize the nutrients present within the concentrate. Whole-food nutrients are recognized by the body as food, and therefore are better utilized and retained by the body when compared to any other supplemental form. Food is the complex form in which all nutrients are found in nature.

Supplements in this category will list the nutrient names--vitamin C, folate, calcium, etc.--and the mg. dosage of each, butthey will not denote a fractionated chemical name. Nutrients as they exist in food cannot be listed as one identifiable organic compound in this way, for the reason that they always exist within an interconnected weave of parts, and never as one isolated element. Nutrients in a whole-food state will be listed with their food source in the supplement facts panel such as vitamin C.

Bioavailability Spectrum

Let's take a closer look at the bioavailability of supplements on a spectrum, from least bioavailable to mostbioavailable. On the low end of the spectrum you would have an inexpensive, store bought, hard-packed tablet. Something brightly Coloured that you can't break with your fingers. Such supplements have many fillers and binders, including ingredients that can be found in wood putty and tar. You can't take these on an empty stomach, or else you'll get an upset stomach.

These are difficult to digest and breakdown, and often times your body doesn't at all. And these have been known to have been passed whole! For that reason you can understand why they have a bioavailability of 0-25%. So, just as a reminder, when you look at the daily values of nutrients on the label of these products, your body is not getting those Percentages of nutrition, but rather up to 25% tops.

Moving up the bioavailability spectrum we come to the capsule delivery system. They have no fillers or binders, and you usually have to take larger quantities of these because they are not very concentrated. The bioavailability is a little better, but still only 40-60%.

Getting higher up the bioavailability spectrum brings us to liquid supplements. These are better then the previous two in bioavailability and range between 80-90%, but many liquid herbs and oils do not taste very good. Finally we arrive to isotonic-capable supplements. Isotonic supplements are the most bioavailable delivery of supplements and are up to 95% absorbable.

Isotonic-Capable Supplements

It is the proper ratio of nutrients to water so that it can be seamlessly integrated into the blood stream. An isotonic capable supplement works similarly. It is the proper measurement of nutrients to water to create an isotonic capable solution. When it is taken on an empty stomach, It is recognized as the same pressure as the stomach. The pyloric valve is signaled to open, and the full concentration of nutrients is delivered to the small intestine (our absorption site) where it is up to 95% absorbable within 5-15 minutes of consumption.

It is important to note that not all liquids are isotonic capable, again it is a specific ratio of medicine or nutrients to water which makes it isotonic. Everything that you ingest and put into your stomach has to be digested down to an isotonic state before it is released to the small intestine. When you eat some food and take a standard pill or capsule vitamin, it all remains in your stomach digesting for anywhere from 40 minutes to 4 hours. At that point a certain amount of dilution of nutrients has taken place. Some low quality pills do not even break down due to harsh fillers and binders, but still pass to the small intestine where none of the content of the supplement can be absorbed. This is why the bioavailability of pills and capsules are so significantly lower than isotonic capable supplements.

Vitamin-A

Vitamin-A was the first fat-soluble vitamin to be discovered. In 1919, it was recognized that egg yolk contained a fat-soluble substance essential for life. In 1924, it was demonstrated that the eye disease, xerophthalmia, in children could be prevented by feeding them butter or cod liver oil. Vitamin-A was purified--separated from its natural constituents--and its refined chemical structure proposed in 1931. Synthesis, an artificial chemically-pure duplication, was accomplished in 1947 at Hoffman-LaRoche, a pharmaceutical company in Switzerland. Vitamin-A is actually a whole family of compounds, including retinol, retinal, retinoids, carotenes, and carotenoids, and always occurs in nature with synergists such as fatty acids, chlorophyll, other vitamins, enzymes, minerals, and trace elements. Retinol, retinal, and retinoids are found in foods of animal origin such as eggs, liver, fish, butter, and cheese.

The scientific term for vitamin-A is retinol, because of its presence in the retina of the eye. The role of retinol in vision was elucidated by a number of brilliant scientists, beginning in 1877 with a German, W. Kuhne, who discovered that the purple retinas from dark-adapted frogs turned yellow when exposed to light. The purple Colour is restored in a complex biochemical cycle involving vitamin-A, which makes vision possible. Other scientists demonstrated the role of vitamin A in cell differentiation, bone development, reproduction and immune system function. Dr. Weston Price confirmed the value of vitamin-A in traditional diets during his studies of primitive peoples carried out during the 1930s and 1940s. Due to the outstanding scientific work of these and many other researchers, the administration of cod liver oil to growing children--a tradition found among Arctic peoples such as the Scandinavians and Eskimos--became standard practice until after the Second World War.

In recent decades, much vitamin-A research has focused on its role in preventing cancer, and its use in combination with nontoxic therapies in the treatment of cancer. Dr. Max Gerson treated many cases of terminal cancer with excellent results, using raw liver juice, a rich source of vitamin-A. Modern confinement farming practices effectively prevent vitamin-A from incorporation into animal foods and the processing industry would rather use vegetable oils than animal fats. Some vegetable oils contain carotenes but they do not contain true vitamin-A. Only animal fats contain vitamin-A and vitamin-A is present in large amounts only when the animals have a source of carotenes or vitamin-A in the diet, such as green pasture, insects, and fish meal.

Many popular books on nutrition insist that humans can obtain vitamin-A from fruits and vegetables. Even worse, FDA regulations allow food processors to label carotenes as vitamin-A. Vitamin-A from butter may be two or three times as potent, unit for unit, as vitamin-A from cod liver oil in some people.

The food industry, and the low-fat school of nutrition that the industry has spawned, benefit greatly from the fact that the public has only vague notions about vitamin-A. In fact, most of the foods that provide large amounts of vitamin-A--butter, egg yolks, liver, organ meats and shellfish--have been demonized. Under optimal conditions, humans can indeed convert carotenes to vitamin-A. This occurs in the upper intestinal tract by the action of bile salts and fat-splitting enzymes.

Of the entire family of carotenes, beta-carotene is most easily converted to vitamin-A. The accepted ratio of how many units of beta-carotene is needed to produce one unit of vitamin-A is 6:1 and some research suggests an even higher ratio. This means that you have to eat an awful lot of vegetables and fruits to obtain even the daily minimal conversion. But the transformation of carotene to retinol is rarely optimal. Strenuous physical exercise, periods of physical growth, pregnancy, lactation and infection are stresses that quickly deplete vitamin-A stores.

Diabetics and those with poor thyroid function, which could include at least half the adult U.S. population, cannot make the conversion. In fact, the thyroid gland requires more vitamin-A than the other glands, and cannot function without it (mercury toxicity diminishes the thyroid and pituitary glands' functions). And a diet rich in vitamin-A will help protect the diabetic from the degenerative conditions associated with the disease, such as problems with the retina and with healing (mercury binds with the sulfur in insulin).

Children make the conversion very poorly and infants not at all--they must obtain their precious stores of vitamin-A from animal fats--yet the low-fat diet is often recommended for children. Vitamin-A from animal sources (retinol) is primarily absorbed within three to five hours after ingestion in the upper intestinal tract; it is here that the fat-splitting enzymes and bile salts convert carotenes into usable nutrients, though conversion and assimilation takes up to six or seven hours. Carotenes are converted by the action of bile salts (mercury inhibits production of taurocholic acid, a bile salt) and very little bile reaches the intestine when a meal is low in fat.

The conversion of carotenes is stimulated by thyroxine, a hormone secreted by the thyroid gland. Zinc, iron, and vitamin-E status also influences carotene conversion. Once converted, the carotenes are absorbed in the same way as retinol and allocated by the liver. Unabsorbed carotene is excreted in the feces.

Butterfat stimulates the secretion of bile needed to convert carotenes from vegetables into vitamin-A, and at the same time supplies very easily absorbed true vitamin-A. Polyunsaturated oils also stimulate the secretion of bile salts but can cause rapid destruction of carotene unless antioxidants are present. It is very unwise, therefore, to depend on plant sources for vitamin-A. This vital nutrient is needed for the growth and repair of body tissues; it helps protect mucous membranes of the mouth, nose, throat and lungs; it prompts the secretion of gastric juices necessary for proper digestion of protein; it helps to build strong bones and teeth and rich blood; it is essential for good eyesight; it aids in the production of RNA; and contributes to the health of the immune system.

Strenuous physical exercise, excessive consumption of alcohol, excessive consumption of iron (especially from"fortified" white flour and breakfast cereal), use of a number of popular drugs, excessive consumption of polyunsaturated fatty acids, zinc deficiency and even cold weather can hinder the conversion of carotenes to vitamin-A, as does the low-fat diet.

Natural vitamin-A provided by liver, eggs, butter, cream and cod liver oil is well recognized as providing excellent protection against birth defects. Vitamin-A deficiency in pregnant mothers results in offspring with eye defects, hydrocephalus, displaced kidneys, harelip, cleft palate and major malformations of the heart and large blood vessels. Vitamin-A stores are rapidly depleted during exercise, fever and periods of stress. Vitamin-A deficiencies are widespread and contribute to high infant mortality, blindness, stunting, bone deformities and susceptibility to infection.

These occur even in communities that have access to plentiful carotenes in vegetables and fruits. It is required for cellular differentiation (determines the function that cells will have): this assures that the cells which are lost from natural turnover, stress, insult, injury, disease, etc. are reproduced in the exact same form as the ones being replaced. Growing children actually benefit from a diet that contains considerably more calories as fat than as protein. Generous amounts of vitamin-A insure healthy reproduction and offspring with attractive wide faces, straight teeth and strong sturdy bodies.

This vitamin assists in normal pregnancy, embryonic development, successful reproduction, fertility, lactation (nursing), and reproductive organ function, spermatogenesis, to adrenal, thyroid, and other gland functions. Vitamin-A is important for normal eyesight; necessary for night vision. A high-fat diet that is rich in vitamin-A will result in steady, even growth, a sturdy physique and high immunity to illness. Vitamin-A-rich foods like liver, egg yolk, cream and shellfish confer resistance to infectious diseases in children and prevent cancer in adults.

Children with measles rapidly use up vitamin-A, which can result in irreversible blindness. An interval of three years between pregnancies, allows mothers to rebuild vitamin-A stores so that subsequent children will not suffer diminished vitality. Kwashiorkor is as much a disease of vitamin-A deficiency, leading to impaired protein absorption, as it is a result of absence of protein in the diet. High-protein, low-fat diets are especially dangerous because protein consumption rapidly depletes vitamin-A stores. Children brought up on high-protein, low-fat diets often experience rapid growth.

The results--tall, myopic, lanky individuals with crowded teeth, and poor bone structure, are commonplace in America. High-protein, low-fat diets can even cause blindness. Scarcity of good quality raw dairy products, a rejection of organ meats as old fashioned or unhealthful, and a substitution of vegetable oil for animal fat in cooking all contribute to the physical degeneration and suffering of Third World peoples. Supplies of vitamin-A are so vital to the human organism that mankind is able to store large quantities of it in the liver and other organs. Even people who can efficiently convert carotenes to vitamin-A cannot quickly and adequately replenish vitamin-A stores from plant foods.

Many factors interfere with its absorption and utilization. Inadequate fat in the diet, poor production of bile salts, low enzyme status, and compromised liver function can all interfere with the uptake and usage of vitamin-A, especially when given as a supplement in the form of retinol, rather than as a component of whole foods. One tablespoon of cod liver oil contains at least 15,000 IU and one serving of liver (should be from organically raised animals) can contain up to 40,000 IU vitamin-A.

Foods high in vitamin-A are especially important for diabetics and those suffering from thyroid conditions. Weston Price considered the fat-soluble vitamins, especially vitamin-A, to be the catalysts on which all other biological processes depend. Efficient mineral uptake and utilization of water-soluble vitamins require sufficient vitamin-A in the diet.

Synthetic Vitamin-A

Synthetic vitamin-A interferes with the proper utilization of natural vitamin-A from foods. Pure retinol is added to many fabricated foods like margarine, white flour, extruded breakfast cereals and pizza. A synthetic retinoic acid derivative,isotretinoin (Accutane) is used to treat acne as well as psoriasis, cancer, and oral leukoplakia. Children whose mothers took Accutane during pregnancy were born with nervous system defects, brain damage, congenital heart disease, and more. Depression, crying spells, malaise, and forgetfulness are reported human symptoms.

Inflammation of the lips (cheilitis) develops in 90% of the users; conjunctivitis develops in 40%. Muscle damage has also been added to the list. The synovial membranes in joints are supported by vitamin-A.

The vitamin-A complex is supportive to the adrenal glands which produce cortisone and other needed hormones; and vitamin-A plays an important role in any inflammation and repair process. Vitamin-A is important to the external and internal mucous membrane tissues as well as to the immune system. However, synthetic retinoic acid metabolites do not reveal any specific supportive mechanism and results are not seen with synthetic, altered fractions.

The whole vitamin-A complex is known to be very functional, but when distilled or synthetic fractions are used, roles and results are questionable at best. Megadoses of synthetic vitamin-A can result in hypertrophy (excessive growth) of the liver and spleen, with elevation of AST (aspartate aminotransferase), a liver enzyme. Liver fibrosis (abnormal formation of fibrous tissue), fat deposition, obstruction of portal blood flow with portal hypertension (high blood pressure) and sclerosis(hardening) of the blood vessels have been found at liver biopsies.

Excess synthetic vitamin-A is also associated with calcium deposits in soft tissues, contributes to attacks of gouty arthritis, hair loss, irregular menses, emotional lability, insomnia, restlessness, night sweats, and more. Vitamin-A is essential to spermatogenesis, the production of sperm in men, and it requires alcohol dehydrogenase, an enzyme to convert retinol to retinal. This enzyme may require the trace element, zinc, as well. Excessive doses of synthetic fractions of vitamin-A can convert to a compound that can be more damaging than a deficiency in its effect on the germinal epithelium of the testes.

Vitamin-A is present in almost twice the amount in seminal fluid of highly fertile men compared with that of men deficient in sperm. Zinc increases vitamin-A uptake by the seminiferous tubules of the testes. Vitamin-A is bound to a protein, called retinol-binding protein.

The necessity of enzymes, trace mineral activators, amino acids, and other constituents stresses the point that the whole food complex is important and effective whereas an isolated portion cannot produce the same results. Excessive fractionated"vitamins" A and E may produce testicular hemorrhaging. Zinc is required to maintain normal concentrations of vitamin-A in the blood plasma.

Diabetic persons tend to have decreased levels of zinc. Since a zinc deficiency may lead to difficulties mobilizing vitamin-A from the liver for use elsewhere in the body, vitamin-A deficiency will worsen the condition. Another essential constituent is fat, as vitamin-A occurs in foods with fatty acids and is a fat-soluble vitamin. It is absorbed much better when natural, unaltered fats are also present. Other fat-soluble vitamins occur with and work with whole vitamin-A complex such as vitamin E complex and essential fatty acids.

When vitamin E is deficient, one develops low blood levels of vitamin-A; when vitamin E is restored to the diet, vitamin-A levels become normal. Yet, using chemically-pure fractions of"vitamins" A and E in large doses creates more problems, more imbalances. Fractionated alpha-tocopherol (so-called vitamin-E) can supplant vitamin A in the body, reducing its effectiveness. Administration of the whole food complex brings excellent responses.

Stubborn cases of gastrointestinal irritation and even ulceration in the digestive tract that failed to respond to other treatments have totally resolved after use of real vitamin-A complex. Persons with cystitis (inflammation of the bladder) experience relief after a relatively short time taking food concentrates of Vitamin-A and other supportive nutritional complexes. Edema caused by liver pathology is another instance where vitamin-A complex is of particular value; the liver is the chief location of vitamin-A stores. Colds, chronic rhinitis, sinusitis, psoriasis, and many other conditions--especially those involving epithelial tissues--are responsive to the whole vitamin-A complex.

A new strategy that has been the development of and promotion of genetically engineering rice to produce carotenes. Golden rice containing carotenes can't provide true vitamin-A to the world's children but it will further the trend of pushing their parents off the farm and into ghastly slums. The most effective response that consumers have is to boycott processed foods!

Vitamin B-Complex

The B-COMPLEX factors naturally divide themselves into two groups. One group consists of the factors that are alcohol and water soluble. This includes B1, B12, Pantothenic Acid and Adenine, but the primary factor in this group is called Vitamin B4, an anti-paralysis vitamin. It is one that has not yet been recognized by the FDA as being necessary in human nutrition. They say it helps pigeons, but not people. So vitamin B4 is a nerve-promoting factor, it promotes the transmission of nerve impulses. This would be needed with heart conditions involving irregular heartbeat, fibrillation, or heart blocks.

The B4 factor also causes vessels to constrict. It's very useful for people with low blood pressure--the person whose blood vessels are just hanging open, with no tone. The B4 puts strength in the vessels and snaps them into action by improving the transmission of nerve impulses to the vessels. For this reason, vitamin B would be contraindicated in cases of organic heart trouble, such as angina. Here you would need the G factor. The G factors are not soluble in absolute alcohol. Complex G contains the lipotropic factors that metabolize fat.

Complex-G contains B2, B3, B6, Choline, Betaine, Inositol, Folic Acid, Biotin and PABA, and is a vasodilator--it opens up the blood vessels by relaxing the nerves in the vessels. So complex G is for a person with high blood pressurethe tense, coronary type person. It is important to have these factors separated and useable either by themselves or together. This benefit is not achieved with megadoses of synthetic so-called

Vitamin-B. Today, the average practitioner of the medical arts, if sufficiently brain washed by the medical-drug-food-dairy-chemical combine of disinformation, cannot conceive that any person could suffer from B-complex deficiency.

Even if they know that refining and processing of food, chemical fertilizers and food additives, all diminish vitamin content and nutrient quality. Because these"foods" are"enriched" with synthetic vitamins. University biochemists and leading health authorities parrot the phrase:"There is no difference between natural and synthetic vitamins." Below are diagrams of the structural formulae of NATURAL vitamin B1 (left) and synthetic B1 (right).

Show any high-school chemistry student with a passing grade the diagrams and tell them they are the same, and that the body does not know the difference. They will laugh in your face. Now, leading nutritionists from the university fountainheads of knowledge will tell you:"there is no difference."

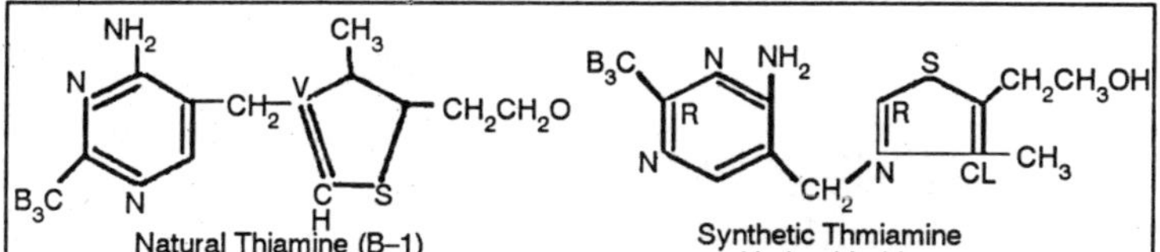

Natural Thiamine (B-1)

Synthetic Thmiamine Hydrochloride

Thiamine is one of two vitamins in the body that contains a sulfur entity; the other is biotin. The high binding affinity that mercury has for the sulfur-containing molecules (thiols and disulfides) gives it the biochemical potential to affect the functions of this critical vitamin and affect critical energy processes in the body. Not only does thiamine have a sulfur atom in its structure to start with, but its involvement as a coenzyme in critical biochemical processes is connected through transfers involving sulfur molecules.

Out of about two-dozen substances produced by the metabolism of thiamine in the body and excreted in the urine, science has only identified the function of six of them thus far. There is a lot going on metabolically with vitamin B1, that mercury can affect, that haven't even been considered as potential problems because the complete biochemical function of the vitamin has not been identified.

Thiamine is involved in many enzyme reactions in the body. As a coenzyme, B1 is involved in the intermediary metabolism of carbohydrates in all cells of the body. For natural vitamin B1 to perform as a coenzyme, it must be phosphorylated or combined with two molecules of phosphoric acid. As such, it then functions as thiamine pyrophosphate (TPP).

In the biochemically active form, it is also called diphosphothiamine (DPT). Better textbooks of biochemistry use the TPP designation as preferable. Early symptoms of thiamine deficiency, are intriguingly similar to mercury toxicity and are non-specific and hard to pinpoint: fatigue, anorexia, weight loss, gradual loss of muscle strength, peripheral neuritis (numness or increased sensitivity, tingling in the extremities, loss of reflexes), irritability, confusion, depression, lack of initiative, poor memory, gastrointestinal problems (including constipation), and low body temperature. Cardiovascular symptoms can include edema of the ankles, feet and legs, palpitation, tachycardia, abnormal electrocardiograms, decreased blood pressure, and difficulty in breathing after exertion.

Thiamine participates in the metabolism of tryptophan to form nicotinic acid. Because of this, a deficiency of thiamine can cause a deficiency of vitamin B6, which is also required in the metabolic pathway of tryptophan. Thiamine is a cofactor in steroid hydroxylation, fatty acid synthesis, and glucose metabolism in the pentose phosphate cycle, which is another energy system. Mercury inhibits glucose-6-phosphatase, a key enzyme in the glucose cycle. The proportion of glucose metabolized by this route is high in the lactating mammary gland, the adrenal cortex, leukocytes and erythrocytes. Each one of the mentioned also are affected by mercury in ways other than glucose metabolism. Not only mercury, but arsenic and lead impact on the biochemical pathways of these vitamins and minerals. If a person is deficient in thiamine or B1, most likely they are also lacking in all members of the B-complex.

Symptoms Include

1 Weakness and fatigue, insomnia, abnormal fears
2 Indigestion (hypochlorhydria), dizziness, rage
3 Decreased appetite, and hostility nervousness,
4 Craving for sweets, instability, depression
5 Neuralgia to neuritis, forgetfulness, and anxiety
6 Muscular soreness, vague fears, apprehension
7 Headache, uneasiness

Vitamin B deficiency leads to heart block and a remarkably slow pulse... Vitamin B deficiency is known to produce a loss of appetite through its effect of causing a loss of tone of the stomach muscles. This loss of tone also may affect the heart and muscular walls of the arterioles--resulting in heart weakness and edema, respectively.

Albert L. Lehninger, in Principles Of Biochemistry, teaches that while muscle cells can oxidize fats for prolonged energy demand and sugar (glucose) for immediate energy...the brain cell has only one Currency For Energy, and that is glucose. Thus, under stress, exaggerated mental activity requires additional glucose for the brain cell to function.

Not only must the brain cell have extra glucose to cope with stress, the cytoplasm and mitochondria of brain cells (neurons), must have all of the members of the B-complex in their biologically-active-form the coenzymes required for anaerobic and aerobic glycolysis. Minerals and trace elements must also be available optimally. Phosphorus, Potassium, Magnesium, Manganese and Calcium play important roles in the conversion of sugar into energy, within all cells of the body. Being deficient in the B-complex factors, one cannot properly convert sugar or glucose to pyruvic acid for the anaerobic glycolytic pathway of sugar metabolism in the cell cytoplasm. Without B1, pyruvic acid cannot be converted to coenzyme A and enter the mitochondria to participate in the citric acid cycle and form ATP. As a result, pyruvic acid accumulates in nervous tissue and piles up on nerve endings causing irritation. Then the cells signal for help.

The adrenal medula comes to the rescue, increases its function and hormone output...and, as a result, provides the mechanism and the stimulus for producing muscle and brain energy by gluconeogenesis.

As long as adrenal hormone output is sufficient, one can survive all the symptoms of B-complex deficiency. The adrenal glands need protein, minerals and vitamin C (not ascorbic acid), especially tyrosinase (organic copper), to produce catecholamines. Without that extra energy, the person will flee, or cry, or hide. Emotionally they usually hide or separate from reality...the classical nervous breakdown, or nervous exhaustion. When this inevitable point is near at hand, the person not only needs total nutritional support, but also adrenal support.

Besides those nutrients mentioned in the glycolytic phase, the Krebs energy cycle uses B3 five times, B1 two times, Pantothenic acid two times and Biotin three times. Biotin, normally synthesized in the intestines, is needed to transport CO_2 from the cells into the circulation.

Heavy metal toxicity blocks Biotin function, while antibiotics and chlorine in water, stop or restrict Biotin production in the intestinal tract. In the anaerobic phase of glycolysis, Fluoride blocks the very important enolase reaction, which impairs glucose metabolism. B2, Riboflavin or"G" combines with several enzymes and enzyme systems forming many flavo-proteins. It is present in retinal pigment contributing to visual acuity. Deficiency signs and symptoms include:

1 Corneal vascularization (red eye)
2 Corneal cloudiness and ulceration
3 Cataracts
legs
4 Photophobia (burning and itching of eyes)
5 Abnormal pigmentation of iris

8 Myelin degeneration
9 Incoordination
10 Loss of strength in arms and
11 Central neuritis
12 Fissures at corners of the mouth

6 Stomatitis with lesions nose and ears	13 Scaliness and greasiness at
7 Glossitis formation	14 Impaired red blood cell

B_3, or Niacinamide. The acid niacin or nicotinic acid as found in plants is a vasodilator...the amide form as found in animals is not. B3 deficiency signs

1 Erythematous cutaneous lesions aggravated by sunlight	10 Failing vision
2 Anorexia, nausea and vomiting	11 Excessive salivation
3 Dilation of blood vessels (aggravated by Niacin)	12 Burning hands and feet
4 Epithelial atrophy of tongue	13 Pain in calves
5 Headaches, dizziness, insomnia, depression, impaired memory of extremities	14 Numbness and weakness
6 Enteritis to ulcers	15 Difficulty in walking
7 Fatty infiltration of liver	16 Absent knee jerk
8 Watery to bloody stools mania	17 Psychoneurosis to stupor or
9 Myelin degeneration (motor and sensory)	

B6, or pyridoxine as a pyridoxal-5-phosphate, functions as a coenzyme in the metabolism of amino acids, fats and fatty acids. Transports amino acids and potassium into cells. Deficiency signs and symptoms include:

1 Seborrheic dermatitis around eyes, eyebrows and angles of confusion	5 Lethargy and mental mouth
2 Glossitis and conjunctivitis	6 Decreased lymphocytes
3 Severe sensory neuritis a. Numbness and tingling in hands and feet	
b. Hyperesthesia	7 Increased Eosinophils
4 Anorexia, nausea, and vomiting	8 Elevated B.U.N. (Blood Urea Nitrogen)

FOLIC ACID

Folic acid deficiency is one fo the most common vitamin deficiencies. Many of the symptoms are similar to those of B_{12}deficiency. As much as 100% may be lost if foods are improperly stored, cooking water is discarded, or foods are reheated or overcooked. Many medications, including aspirin and anticonvulsants, may also interfere with folacin absorption and metabolism, further decreasing its availability. Folic acid is involved in the metabolism of pantothentic acid and is required for its utilization.

It is through this participation that it is involved in antibody production of the immune system. Folic acid also has a relationship with other vitamins including vitamin C and E. In the body, folic acid is rapidly converted to the biologically active form tetrahydrofolic acid (THFA) in the presence of NADPH (niacin's coenzyme form) and ascorbic acid. In conjunction with vitamin B_{12}, THFA participates in amino acid conversions and the methylation of choline, methionine, serine, and histidine. Folates carry out their metabolic function as a carrier of one-carbon units in the tetra-hydro form. It is this capability to accept one-carbon units, which synthesize various coenzymes, that makes folates so important and involves them in so many metabolic functions.

The liver actively reduces and methylates folates which are then transported into the bile for reabsorption by the gut and subsequent delivery to the tissues. This important pathway provides as much as 200 micrograms or more of folate each day for recirculation to tissues. The folates are a family of coenzymes that function in association with their respective enzymes to accomplish many intracellular metabolic functions. Methyl-THFA is required as the methyl donor in the conversion of homocysteine to methionine (methyl methionine).

This reaction utilizes vitamin B_{12} as a cofactor and also requires vitamin B_6. Folates are also involved in the conversion of the amino acid serine to glycine, and histidine to glutamic acid. It is also necessary in the synthesis of glucose, and DNA.

Folic acid and vitamin B_{12} are both involved in synthesizing DNA. In megaloblastic anemia (folate deficiency), certain steps in this vital process of DNA synthesis are reduced, causing an arresting of the synthesis of red blood cell replicaiton. The form of folacin most commonly found in the liver and serum is methyl-THFA. Methyl-THFA can only recycle through a B_{12}-dependent pathway. If a B_{12} deficiency exists, folic acid is trapped as methyl-THFA and is useless to the body. A deficiency of either vitamin B_{12} or folic acid will result in identical hematological symptoms.

A vitamin B_{12} deficiency impairs the conversion of homocysteine to methionine. The enzyme methionine synthetase requires vitamin B_{12}. Therefore, failure of methionine synthesis leads to inadequate formulation of THFA. Methionine deficiency has been equated with B_{12}deficiency. The phagocytic function of polymorphonuclear leucocytes (PMNs), and to a lesser degree bactericidal function, are adversely affected by folic acid deprivation. Folic acid requirements increase during pregnancy, and low tissue-folate levels are a predisposing factor in the bacteriuria (bacteria in the urine) of pregnancy. There is great potential for mercury and lead to be involved in folate deficiency. The involvement of chronic inhalation of mercury Vapour from mercury amalgam dental fillings can be an etiological factor in thyroid dysfunction involving folic acid, B_{12}, and thyroxine production.

PANTOTHENIC ACID

The physiological active form of pantothenic acid is coenzyme A, which is required for many different enzymatic actions. It is at the Centre of energy metabolism and fat, acetylcholine, and antibody synthesis. Coenzyme A, which is synthesized from pantothenic acid, is abbreviated as CoA or CoA-SH, and functions as a transient carrier and a cofactor for a variety of enzyme-catalyzed reactions involving the transfer of acetyl (two carbon) groups. The CoA-SH designation reflects the fact that the coenzyme A molecule has a reactive thiol (-SH) group, to which acyl-groups become covalently linked to formthioesters during acyl-group transfer reactions. All known acyl derivatives of CoA and related pantetheine derivatives are thiol esters. Because of these derivatives, CoA is involved in the metabolism of carbohydrates, lipids, protein, and porphyrin. It is also involved in the synthesis of fatty acids, cholesterol, citrate, acetoacetate, and the neurotransmitter acetylcholine. Pantothenic acid is an essential component of the brain and must enter the brain and cerebrospinal fluid from the blood. In the brain, as part of CoA, pantothenic acid is involved in many important reactions.

Symptoms identified with pantothenic acid deficiency such as fatigue, headache, insomnia, nausea, abdominal cramps, occasional vomiting, paresthesias of the hands and feet (burning foot syndrome), muscle cramps, and impaired coordination. Pantothenic acid is required for the synthesis of cholesterol precursors for sterol hormones in the adrenal cortex, and a pantothenic deficiency can produce cortical necrosis. Conversely, pantothenic acid stimulates the adrenal glands and increases production of adrenal hormones. Mercury can inhibit production of adrenal hormones and cause adrenal over-growth.

Mercury causes a direct defect in adrenal steroid biosynthesis with physiological consequences being lowered plasma levels of corticosterone and elevated concentrations of progesterone anddehydroepiandrosterone, all abnormal steroid hormone profiles. Mercury's affinity for thiols, sulfhydryl groups and the sulfur molecule, wherever they are biochemically available in the body, gives it the potential to affect countless metabolic functions. Symptoms produced by a pantothenic deficiency are all also documented as effects produced by mercury toxicity.

Vitamin B6 (Pyridoxine)

Vitamin B6 participates in the metabolism of protein, carbohydrates, and lipids, and is a coenzyme constituent in the formation of erythrocytes and in amino acid metabolism. The conversion of methionine to cysteine is dependent on vitamin B_6. The three forms of vitamin B_6 that occur in foods are pyridoxine, pyridoxal phosphate, and pyridoxamine phosphate. The physiologically active forms of vitamin B_6 are pyridoxal phosphate and pyridoxamine phosphate. All three forms are converted in the body to pyridoxal phosphate. Pyridoxal phosphate is involved with a number of enzymes in catalyzing reactions of amino acids.

The most common of these are called transaminations, in which an amino group is transferred to a carbon atom. In these reactions, pyridoxal phosphate serves as a transient carrier of the amino group from its donor to the amino group acceptor. The derivative of this intermediate action is pyridoxamine phosphate which then donates its amino group to the carbon atom of alpha-keto acid, after which it reverts to its pyridoxal phosphate form. In the biochemical reactions related to the conversion of methionine to cysteine, there is a need to transfer a sulfur component, and an amino component, which would make the process doubly vulnerable to mercury because its highest binding affinity is for the sulfhydryl group and then the amino group. Further, mercury can bind to phosphate groups, which gives it a total of three possible mechanisms to affect the metabolism of pyridoxal phosphate, the biologically active form of vitamin B_6.

One of the functions of vitamin B_6 is called transsulfuration. This is the process whereby one sulfur molecule can replace another and represents an important biochemical function for utilization of sulfur. When pyridoxine is deficient in the body, methionine metabolism is altered. The pancreas is very active in transsulfuration and is known to have a rapid uptake of methionine, because of this, mercury can be involved in pancreatic dysfunction. Cystathionine is a major constituent of the pool of free amino acids in the brain. This fact raises the question of neurochemical involvement. Although vitamin B_6 is not synthesized in the brain, it readily enters the cerebral spinal fluid and brain from the plasma. Once within the CSF, B_6 can enter brain cells. The holoenzyme needed to convert cystathionine to cysteine contains B_6 in its pyridoxal phosphate form. It is relatively easy to deplete brain

B6. Any substance that impedes or inhibits the metabolic functions of vitamin B_6 in humans could have some very serious consequences. Symptoms of vitamin B_6 deficiency, like those of micromercurialism, can be nonspecific and hard to pinpoint. They include weakness, mental confusion, irritability and nervousness, insomnia, poor coordination in walking, hyperactivity, convulsions, abnormal electroencephalogram, declining blood lymphocytes and white blood cells, anemia, and skin lesions. Mercury has the biochemical capability of reducing the availability of, or inhibiting the function of vitamin B_6. There is a positive correlation between brain mercury levels and the numbers and surfaces of amalgam fillings.

Vitamin B12

Vitamin B_{12} has the most complex structure of all vitamins and is also unique in that it is the only vitamin containing a metal ion. Cobalamin is the generic name of the vitamin because of the presence of the metal ion cobalt in its structure. Vitamin B_{12} was isolated in 1948. Merck and Company is the holder of various patents of procedures for producing synthetic $B_{12}$and owns the product claim. B_{12} functions primarily as an essential coenzyme for the normal metabolism of all cells, including those of the gastrointestinal tract, bone marrow, and nervous tissue. It is also a coenzyme in the synthesis of red blood cells and in the maintenance of nerve cells, and is involved with protein, lipid and nucleic acid synthesis. It is considered necessary for growth. The coenzyme of vitamin B_{12} is a carrier of methyl groups and hydrogen and is necessary for protein, fat, and carbohydrate metabolism. Coenzyme B_{12}, methylcobalamin, functions as a methyl-group donor to form methionine from homocysteine. Vitamin B_{12} coenzyme deoxyadenosylco-balamin functions in the conversion of methylmalonic acid to succinic acid. Methionine and B_{12} coordinate in the regulation of folate metabolism. A deficiency of either vitamin results in defective synthesis of DNA in any cell that attempts chromosomal replication and division.

Chronic mercury inhalation from mercury amalgam dental fillings, with its great affinity to bind to methionine and cysteine has the potential to decrease the availability of these amino acids and affect the metabolism of both vitamin B_{12} and folate in man. Methionine is needed in choline synthesis, which means that vitamin B_{12} plays a secondary role in this lipid pathway. A choline deficiency that causes fatty liver can be prevented by vitamin B_{12} or the other methyl donors--betaine, methionine, and folic acid. Impaired fatty acid synthesis, observed in vitamin B_{12} deficiency conditions, can result in impairment of brain and nerve tissue. The insulation around nerve cells, the myelin sheath, is misformed in vitamin B_{12}deficiency, contributing to faulty nerve transmission.

Neurological disturbances result from prolonged vitamin B_{12} deficiency. Inorganic mercury inhibits outgrowth of nerve Fibres and the development of glial cells, and also depresses the outgrowth of fibroblasts. Methyl-B_{12} is able to inhibit the neurotoxic effects of organic mercury but not those caused by inorganic mercury. Mercury Vapour released from dental amalgam fillings is inorganic mercury. Very low concentrations of mercury and arsenic are able to inhibit nerve outgrowth, and lead directly interferes with the process of myelination. Proper DNA replication is dependent on the function of coenzyme vitamin B_{12} as a methyl-group carrier. Methyl-B_{12} may play an important role in immune regulation. Methyl-B12 is able to enhance the activity of helper T-cells for immunoglobulin synthesis of B cells.

The presence of methyl-B12 significantly potentiates the induction of suppressor cells. Polymorphonuclear leukocytes (PMNs) are white blood cells that function in a phagocytic capacity (capable of destroying foreign substances). Low levels of mercury inhibit the metabolic reaction called respiratory burst in human PMNs. Mercury acts as an interference to normal biological functions involving the formation and/or function of sulfur containing proteins in cells, enzymes, and hormones. Biological damage will occur long before the appearance of any clinically observable signs and symptoms of mercurialism. Known causes of B_{12}malabsorption are a lack of intrinsic factor and hydrochloric acid in the stomach, the removal or disease of the second portion of the ileum, and competition for B_{12} by microorganisms or parasites.

There is far more unknown than known about B12 and its metabolism in the body. There are several vegetarian food sources of B_{12}. While these sources may not be constant, it appears that they occur with sufficient frequency to supply the minute quantities of B_{12} that are needed by those who are not abusing their health. These include: wheat, soybeans, various common greens, olives, fruits, and many other foods that occasionally have B_{12} either in or on the food. Vitamin B_{12}has been found in roots and stems of tomatoes, cabbage, celery, kale, broccoli and leeks. It has been found in the leaves of kohlrabi. Seaweed and alfalfa contain B_{12} also. It is difficult to avoid vitamin B_{12}.

A further source of B_{12} is bacterial growth in the mouth (around the teeth and gums), in the nasopharynx, around the tonsils and tonsillar crypts, in the folds at the base of the tongue, and in the upper bronchial tree. Up to 0.5 micorograms daily can be obtained from this source. It's likely that this source alone will supply sufficient quantities of B_{12} for the small requirement that a pure vegetarian has, especially considering his low protein intake, which further reduces the need for B_{12}. It has also been shown that some bacteria, which may colonize the small intestine of man, can synthesize considerable amounts of biologically active forms of the vitamin. The terminal ileum is principally involved in absorbing B_{12}. Since bacteria can form B_{12} above the terminal ileum, and since intrinsic factor, which is needed for most rapid absorption, is found in the ileum, there can be adequate absorption of vitamin B_{12} produced from these bacteria in the ileum in healthy people.

Individuals taking conventional diets need only about 0.1 micrograms of B_{12} per day even though conventional diets contain excess fat, animal protein and refined foods all of which increase the need for B_{12}. True vegetarians require only 0.05 micrograms of B_{12} daily. An ounce of the roots of leeks, beets and other vegetables will provide 0.1 to 0.3 mcg. of B_{1}, more than a day's supply. The routine use of vitamin B_{12} is not advised. Cases of increased cancer production in animals receiving high levels of B_{12} have been published. A group of French investigators reported a series of cases suggesting that B_{12} may stimulate multiplication of cancer cells and aggravate the disease. Patients with rheumatoid arthritis present serum B_{12} levels significantly higher than normal subjects. It is also known to be higher in patients with ulcerative colitis, leukemia and other serious illnesses. Breast-fed infants get only a tiny fraction of a microgram of B_{12} per day, even if their mothers are on a high intake. Yet, they can build neurological tissue, convert their hemoglobin from F to A type, and do other complex functions dependent on B_{12}. Iron deficiency interferes with absorption of B_{12}. Large doses of vitamin B_1 can also destroy vitamin B_{12}, as can the use of oral contraceptives. Lactose intolerance may increase B_{12} needs.

VITAMIN C COMPLEX

Another danger inherent to crystalline-pure, megavitamin usage is the consumption of ascorbic acid--the so-called vitamin C that is recklessly prescribed by the"pseudo-nutritional pharmacologists" (Physicians, Nurses, Nutrition consultants, registered dieticians, health food store employees, etc.), who have little, if any, realistic training in this most important field. Ascorbic acid is not vitamin C. Ascorbic acid is a fraction of the biologically utilizable Vitamin C complex. Most all ascorbic acid sold is synthetically produced from corn sugar and even though it is synthetically manufactured, it is called"natural" and"organic" because corn is found in nature. Technically, anything with a carbon atom can be called"organic." The terms natural and organic are totally meaningless.

This article alerts the general public to the fact that as little as 1,500 milligrams of ASCORBIC ACID, taken daily, depletes blood copper, leading to serious health problems. You jeopardize your important"fight or flight" (stress) response by depletion of copper. Just 500 mg. of ascorbic acid with 3 meals can do just that. Subclinical vitamin C deficiency may exist in a large segment of the population, and can lead to impaired health and increased susceptibility to other diseases. Ascorbic acid therefore is incorrectly named. With isolated ascorbic acid (from Hungarian red pepper), he could not stop the capillary hemorrhaging, so characteristic of scurvy, that he could reliably influence with the C-complex. He went back to the Labouratory and discovered vitamin P, the rutin factor of the C-complex, which exists more in buckwheat (grain and leaf) than in citrus.

The P-factor of the C-complex gives capillary and blood vessel wall strength, the J-factors, increase the oxygen-carrying capacity of the blood. The known physiological functions of vitamin C are synthesis of polysaccharides and collagen; formation of cartilage, dentin, bone, and teeth; antioxidant; absorption of iron; cold tolerance; maintenance of the adrenal cortex; metabolism of tryptophan, phenylalanine, and tyrosine; growth; wound healing; and maintenance of capillaries. Vitamin C is directly involved in proline and lysine hydroxylation; carnitine synthesis; dopamine hydroxylation; drug and cholesterol breakdown; sulphation; lymphocyte and neutrophil function; and folate reduction.

Vitamin C has a protective effect against mercury poisoning. For many years, mercury diuretics were used extensively by the medical profession. The toxicity of these mercurial diuretics could be reduced if the patient was given vitamin C prior to or simultaneously with the mercurial diuretic. When vitamin C is metabolized, part of it is metabolized to vitamin C-sulphate, with the sulphate being derived from sulphur-containing amino acids such as cysteine. Vitamin C competes with certain drugs for sulphate conjugation, which could affect the pharmacological activity and toxicity of drugs.

Mercury competes actively for the sulphur-containing amino acids cysteine and methionine, as well as the cysteine molecule of the tripeptide glutathione. Anyone with amalgam fillings should routinely eat a diet rich in sulphur-containing foods to offset the depletion of cysteine and methionine caused by the inhalation of mercury Vapour being continually released from amalgam fillings. The problem of cysteine depletion is further aggravated by the presence of mercury, lead, arsenic, and cadmium, because they are all thiol sensitive.

Mercury is also known to inhibit collagen synthesis. The synthesis of collagen is impaired in vitamin C deficiency, due to lowered ability to hydroxylate lysine and proline. The reducing agent in the hydroxylation of both lysine and proline is the reduced form of vitamin C. Consequently, in vitamin C deficiency, the amount of effective collagen Fibre present in connective tissue is reduced. In severely scorbutic patients, swelling, hemorrhages, and secondary bacterial infections of the gingival margins are common.

It is not the actual deficiency of vitamin C that causes the inflammation, but the lack of real vitamin C, impairing the normal defensive responses of the mucous membranes. Thus, the massive gingival enlargement so characteristic of scurvy results from the combined effects of lack of vitamin C and nonspecific inflammation. The adrenal glands are one of the target glands of mercury deposition. They also contain the body's second highest tissue levels of vitamin C.

The physical response to stress is an increased secretion of the hormones of the adrenal glands. Physical and mental stress increases adrenal activity which, in turn, depletes vitamin C from the gland. Humans replenish the adrenal stores of vitamin C by taking it from other stores in the body.

If tissue values of vitamin C are low, there may be an insufficient amount available to replenish or satisfy the requirements of the adrenals. Thiol groups have a direct role in the preservation of tissue vitamin C and its reduced form, and any decrease in the biological reducing capacity of tissue thiols will result in lower levels of tisssue vitamin C. Under these conditions, normal adrenal hormone response may become inadequate. Prolonged exposures to low mercury concentrations depress the adrenal vitamin C content.

The person with amalgam dental fillings who is being exposed to chronic intakes of mercury Vapour is subjecting the adrenals to depletion by chemical stress. One function of the liver tissue is to reduce dehydroascorbic acid to ascorbic acid. This ability guarantees the preservation of the body's vitamin C reserves, so necessary to normal function. The body doesn't want ascorbic acid *per se*; it wants the C-complex being preserved by ascorbic acid. Once in the body, the body sheds the ascorbic acid, the preservative wrapper, the way you shed the peel of an orange; takes the rest of the C-complex and places it as the lance of the lymphocyte; and kills bacteria with it. A lymphocyte unarmed with C complex will fail in its function, it will be impotent, it won't be able to destroy infectious organisms. Lymphocyte potency is dependent on copper, the core trace mineral of the tyrosinase enzyme.

The reason one gets any relief from taking ascorbic acid is that ascorbic acid lowers the pH to the acid side of the pH scale. It acidifies the body, creating an unfavourable medium of growth for pathogenic bacteria. Acting in this way, ascorbic acid is not producing a"vitamin effect" but rather an"acidifying effect." Most infectious pathogenic bacteria thrive in an alkaline pH.

The following bacteria, all well-known enemies of modern science's war on bacterial infection, grew optimally on alkaline media of pH 7.4 and above: staphylococcus (staph infection), streptococcus (strep throat), pneumococcus (pneumonia), h. influenza (the flu), meningococcus (meningitis), corymbacterium diptheria (diptheria), clostridium tetani (tetanus), and others.

Apple cider vinegar is 5% malic acid, a natural Colourless crystalline acid normally found in the body. It can be taken with water during meals, and is much less expensive than ascorbic acid and does a better job in lower doses at acidifying the body, without causing rebound deficiencies as ascorbic acid does. Ascorbic acid has long been recognized as a contributing factor in tyrosine (copper) depletion. Tyrosinase is absolutely essential in the biosynthesis of catecholamines.

The adrenal medulla supplies the catecholamines, epinephrine and norepinephrine which are biosynthesized from phenylalanine being oxidized to tyrosine, to dopa (dihydroxyphenylalanine), dopa is transformed first into dopamine and then into norepinephrine with the enzyme tyrosinase catalyzing the intermediate reaction. Other functions of organic copper include: vascular integrity, purine excretion, hemoglobin stability, gas exchange, digestive enzymes, myelin maintenance, visual accommodation, fertility, adrenal hormone synthesis, and connective tissue integrity.

The former, catalase, is an enzyme that cooperates with SOD in the removal of free oxygen radicals from cellular metabolism. Excess quantities of so-called vitamin C can reduce the pH of the urine to as low as 4.0, and cause the precipitation of large quantities of urates, increasing incidence of kidney stones. It appears that excessive doses of ascorbic acid interfere with purine metabolism, increasing the risk of gout. There is some evidence that high doses of ascorbic acid reduce fertility in some women. Large doses of ascorbic acid and vitamin E are undesirable as they may reduce availability of vitamin A.

Ascorbic acid increases intestinal peristalsis and may produce diarrhea. This laxative action, along with its acidifying effect, may be the only benefit ascorbic acid has against a cold. High doses of so-called vitamin C, as popularly used in home remedies for the common cold, destroy substantial amounts of vitamin B12.

Ascorbic acid in quantities in excess of a half a gram will destroy 50% to 95% of B12 content in food. Megadoses of phony vitamin C may produce B12 deficiency by destroying the cobalamins during transport through the digestive tract, and possibly also in the tissues. No further discussion should be needed to establish the lack of wisdom leading to taking or prescribing ascorbic acid.

Vitamin E

The chemical name for vitamin E is tocopherol, which is derived from the Greek tokos (childbirth) and pherin (to bear). The ending ol is the chemical suffix to denote an alcohol. The name tocopherol was bestowed on this vitamin in 1938. Vitamin E is a fat-soluble vitamin, as are vitamins A, D, and K. Fat-soluble vitamins can be stored in the body whereas water-solubles cannot. Fat-soluble vitamins are excreted chiefly by the fecal pathway versus the urinary pathway for the water-soluble vitamins. They are absorbed along with dietary fats, and conditions of extremely low fat intake or impaired uptake of fats will also interfere with their absorption. Antibiotics and certain other drugs, as well as certain disease states such as malabsorption syndromes, decrease the absorption of fat-soluble vitamins from the intestinal tract.

One of the primary biological functions of vitamin E relates to its role as an antioxidant. Vitamin E's antioxidant ability is enhanced by selenium. This synergistic effect preserves membranes from destruction by oxidation products and especially retards breakdown (hemolysis) of red blood cells. Vitamin E can reduce the toxic effects of mercury. Vitamin E is able to reduce the chromosomal breakage caused by mercury. Vitamin E also has sulfhydryl-protective activity.

When vitamin E is involved in reducing the toxic effects of mercury, there is less available for normal metabolic functions. Vitamin E protects the enzymes G6P, G6PD, GSH-Px and glutathione reductase from auto-oxidative damage and at the same time protects the biological supply of glutathione. The presence of mercury in combination with a vitamin E deficiency has the potential of producing toxic reactions in the body. The primary lymphoid organs are major targets of selenium and vitamin E dietary deficiencies, contributing to impaired immune function.

Vitamin E is not a cofactor, such as the trace elements manganese, copper, zinc, and selenium, it is an independent antioxidant that has its main effect in the lipid phase of the cell, protecting the membrane from auto-oxidative damage. A distinguishing feature of cystic fibrosis is fat malabsorption. Cystic fibrosis patients usually have a deficiency of vitamin E. Patients with G6PD deficiency have less capacity to generate reduced glutathione, and their red blood cells have a greatly reduced half-life. Vitamin E supplementaton can decrease the level of neonatalhyperbilirubinemia due to G6PD deficiency. Mercury inhibits G6PD activity and also reduces available vitamin E.

Alpha and delta tocopherols or mixed tocopherols are not vitamin E. They are merely antioxidant or protective agents (banana peel) for the more complex parts of the vitamin E assemblage (xanthine, phospholipids, lipositols, sex hormone precursors)--see functional architecture of vitamin E. There are seven tocopherols in all, alpha being one of them. The government has decided to rate any vitamin E product according to the amount of alpha tocopherol it contains.

The tocopherols are antioxidants, just like ascorbic acid. They are a part of the E complex, but they are not the active ingredient. Their function is to preserve the active factor. Tocopherols are measured on the basis of cell growth. The entire E complex contains the polyunsaturated fatty acids. (Vitamin F) Vitamin F is part of the E complex. Also, vitamins A and K. There are some forms of vitamin D and manganese in the E complex, as well. The traditional source for vitamin E is vegetable oils. But the best quality vitamin E comes from vegetables. Lettuce is one of the best sources and the pea plant is equally as good. If you take any Labouratory-chelated products, your body has to do some combining to make them work. But if you take your minerals the way nature provides them, combined with protein and vitamins through the growing process of plants, they are bound to work. For example, we could take vitamin E and separate the tocopherols from the manganese.

Putting them back together in the Labouratory--chelating them artificially--we would find they wouldn't work like they did before you took them apart. If you take a watch apart piece by piece and Analyse it and then throw all the parts back together, it isn't going to work! Although it is the same watch, it has now lost its function. A nutritional concentrate is an organized mechanism that has a function. If you break the function by separating the parts, the body can no longer use the mechanism. High potency vitamin E is made this way. Tocopherols are broken down and separated from oils--the natural tocopherols are separated from vegetable oil. Then they are packed into a gelatin perle, and you have a high potency vitamin E product--a 220 or 500 International Unit (IU) Vitamin

E product. That means 200 or 500 International Units) of alpha tocohpherol--the potency of the product is measured by the amount of alpha tocopherol it contains. Although the label of such a product will say"all natural" or"organic""--it was taken from oil, a natural source--it's still not food. It's an incomplete product. You're getting lots of tocopherols, but none of the active components of the E complex; the F, A, K, D, manganese, selenium, etc.

The Vitamin E2fraction (prostaglandin) is found in beef chromatin material, which is glandular substance. It's a food factor that will stop a heart cramp or a heart pain, usually within about ten minutes. Most cramps are due to calcium deficiency. So E2 gets calcium into the area to relax the muscle. It can be used to relieve any muscle cramp, but it's especially effective to relax cramps in the muscles of the heart.

Natural forms of vitamin E complex lose up to 99% of their potency when separated from their natural synergists. These synergists include tannins, fatty acids (vitamin F complex), phospholipids, and other synergistic factors. Chemically purified vitamin E (tocopherols), in high unit doses, reverses its effect and produces the same symptoms as a deficiency (including bone decalcification). Without selenium (the trace mineral activator of vitamin E), vitamin E cannot perform many of its functions.

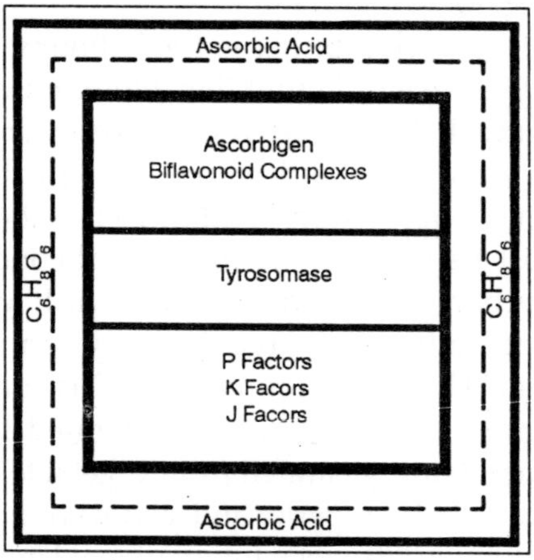

Fig. Function Architecture of Vitamin C-Complex

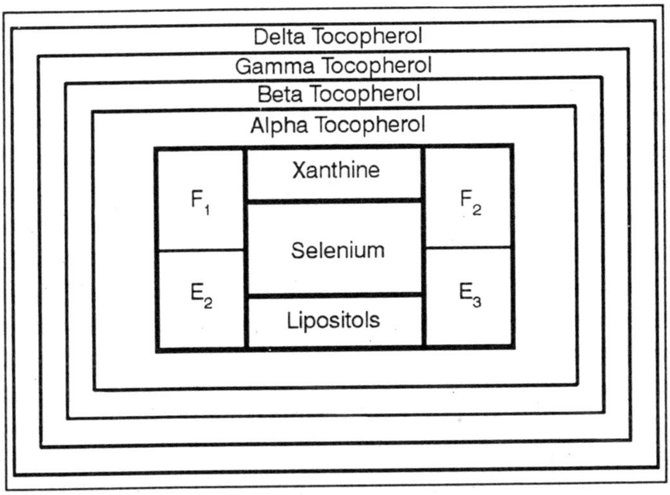

Fig. Functional Architecture of Vitamin E-Complex

Xanthine increases the antioxidative function of the tocopherols as much as 50%. Xanthine plays an important role in the metabolism of purines (nucleoproteins) within the cell. In one instance, hypoxanthine is transformed biochemically to xanthine with the metabolic fate proceeding to uric acid in two forms; enol and keto. Intracellular utilization of purines is dependent upon the said contribution. Lipositols are essential to cholesterol metabolism in order that cholesterol be altered to a biochemical acceptable to the synthesis of adrenal and sex hormones.

F1 is the active form of unsaturated fatty acids described in the Annual Review of Biochemistry, 1949. Vitamin F restores and maintains Calcium in tissue fluids and is anticarcinogenic. F_2 is part of the phospholipid complex in nerve structures; particularly the myelin sheath. It also promotes tissue repair and enhances the benefits of the antiarthritic Wulzen factor. F_2 could very well be the missing link in pernicious anemia according to Dr. Royal Lee. E_2 is a part of E complex that has been compared to nitroglycerine in its function of relieving angina-type pains--the hypoxic spasm of the heart muscle. E_2 reduces pain because it increases the oxygen conserving factor of the blood. The E_3 component of Vitamin E complex is a precursor of sex hormones. Drugs in five major groups have been shown either to function as vitamin B_6 antagonists, or to increase the turnover of vitamin B_6 in the body. Drugs in ten major groups have been shown to affect the absorption of folic acid; to act as folate antagonists; or to increase the turnover or loss of folate from the body. Drugs in four groups have been shown to affect the absorption of vitamin B_{12}. Depletion of one vitamin can affect the requirement for another vitamin. Mercury can cause similar deficiencies or metabolic changes in the way the body handles these same three vitamins.

While man can isolate from foods known factors comprising about 99% plus of total weight, we cannot maintain normal nutrition or"life" by consuming isolated factors. Nature still withholds that"something" that sustains"life." One of the deceptive practices of manufacturers of pseudo-nutrients is to label crystalline-pure and synthetic vitamin fractions as complexes. The meaning of complex is simply: the sum or combination of various things. Another, but more precise, definition is: Consisting of various connected or interwoven parts; composite. As a result, the consumer buys a"vitamin" product that says on the label:"Vitamins From Natural Sources (usually synthetic, from corn sugar and coal tar) with Bioflavonoids or Rutin." Legally, this can be called a"Vitamin Complex" but it is really no more than a mixture of two or more dead, refined chemicals--not a required for life,"biochemical." Remember, the human anatomy is composed of 100 trillion cells that need 75+ nutrients every day and, in turn, it can require six to eight months or more to overcome a nutritional deficiency. If you try and fool mother nature, she'll make a fool out of you.

READING VITAMIN LABELS

NATURAL in this writing means,"As naturally found in nature or natural food sources." It further implies that in preparation there has been no change, either biologically or chemically, in the combination or in the activity of these natural foods. This does not mean the moisture and Fibre cannot be removed.

CO-NATURAL or CRYSTALLINE A co-natural vitamin is one that has a natural food as its source, but has been reduced to one specific pure crystalline vitamin by means of solvents, heat or distillation. Since it is practically impossible to reduce any natural food product to an absolutely pure state during manufacture, the impurities remaining could be certain synergists.

Synthetic A synthetic vitamin is created by a scientist in a Labouratory when he/she has reconstructed the exact molecular structure of a vitamin by putting together or chemically combining molecules from other know sources. The synthetic vitamin has no synergists, no enzymes or co-enzymes, minerals, mineral activators, or co-vitamin helpers. It is a chemically pure vitamin but is, conformationally, a mirror image of the real thing. Legally it is not necessary to give the source from which the synthetic chemical is derived (coal-tar)

Item	If Source is	It is
Vitamin A	Fish oils	Natural
	Acetate	Synthetic
	Palmitate	Synthetic
	Lemongrass	Co-Natural
Table. Vitamin B		
Complex		
Vitamin B_1	Yeast (unenriched)	
	Natural	
(Thiamine)	Thiamine mononitrate	Synthetic
	Thiamine hydrochloride	Synthetic
Vitamin B_2	Yeast (unenriched)	Natural
(riboflavin)	Yeast (unenriched)	
Pantothenic Acid	rice bran, liver	Natural
	Cal panto	Synthetic
Vitamin B_3	Yeast (unenriched),	Natural
	or bran	
(Niacin)	Niacinamide	Co-Natural
Vitamin B_6	Yeast (unenriched)	Natural

(Pyridoxine)	Pyridoxine hydrochloride	Synthetic
	Liver or Yeast (unenriched)	Natural
	Cobalamin from fermentation	Natural
Vitamin B_{12}	Cyano cobalamin	Co-Natural/ Crystalline
	Cyano cobalamin	Co-Natural/ Crystalline
PABA	Yeast (unenriched)	Natural
	Para-aminobenzoic acid	Synthetic
Folic Acid	Yeast (unenriched) or Liver	Natural
	Pteroylglutamic acid	Synthetic
Inositol	Soy beans	Natural
	Reduced from corn	Co-Natural
Choline	Soy beans	Natural
	Choline chloride	Synthetic
	Choline bitartrate	Synthetic
Biotin	Liver	Natural
	d-Biotin	Synthetic
Vitamin B_{15}		
(Pangamic acid)	Bran-apricot kernels	Natural
	Pangamic acid	Synthetic
Vitamin C-Complex		
Vitamin C	Buckwheat leaf, peppers,	
tomatoes, citrus, rose hips	Natural	
	Acerola berries	Co-Natural
	Ascorbic acid	Synthetic
Vitamin D-Complex		
Vitamin D	Fish oil	Natural
	Irradiated ergosterol	Co-Natural
	Calciferol	Synthetic
Vitamin E-Complex		
Vitamin E	Wheat germ oil	Natural
	dl-alpha tocopherol	Synthetic
	d-alpha tocopherol	Synthetic

	acetate	
	d-alpha tocopherol	Synthetic
	succinate	
Vitamin F		
(Unsaturated fatty	Flax oil, pumpkin seeds,	
acids)		
walnuts, soybeans	Natural	
Vitamin K	Alfalfa, Cabbage	Natural
	Menadione	Synthetic

Folic Acid (Pteroylglutamic Acid)

Folic acid is derived from the addition of a pteroyl group to the amino acid glutamate. Folic acid can be reduced to form tetrahydrofolate, which is an important acceptor of one-carbon units. Folic acid plays an important role in the conversion of homocysteine to methionine by providing the methyl to the vitamin B,2 dependent methyl transferase. Therefore, folic acid levels have a strong influence on the endogenous concentration of homocysteine. Elevated levels of homocysteine are associated with increased risk of coronary heart disease. Supplementation with folic acid has been shown to correct plasma homocysteine levels and decrease the morbidity and mortality from atherosclerotic disease.

6

Proteins

PROTEIN PRIMARY STRUCTURE

Proteins are polymers of amino acids.Each amino acid has the same fundamental structure, differing only in the side-chain, designated the R-group.

The carbon atom to which the amino group, carboxyl group, and side chain (R-group) are attached is the alpha carbon (C).

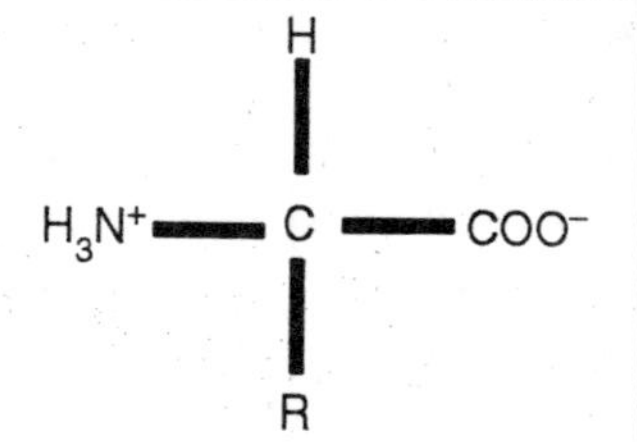

Amino Acid

Structures of amino acids, with their R-groups in blue, are shown below with their three-letter and one-letter abbreviated names. Glycine has just a hydrogen atom in place of an R-group.

At physiological pH, some amino acid R-groups are charged, because of dissociation or association of a proton by, *e.g.*, a carboxyl or amino group. Note that histidine is shown charged, but its pK is close to the physiological range.

Some side-chain groups that are uncharged at the near-neutral pH of the cytosol or extracellular space, may dissociate or gain a proton in the microenvironment of an enzyme active site.

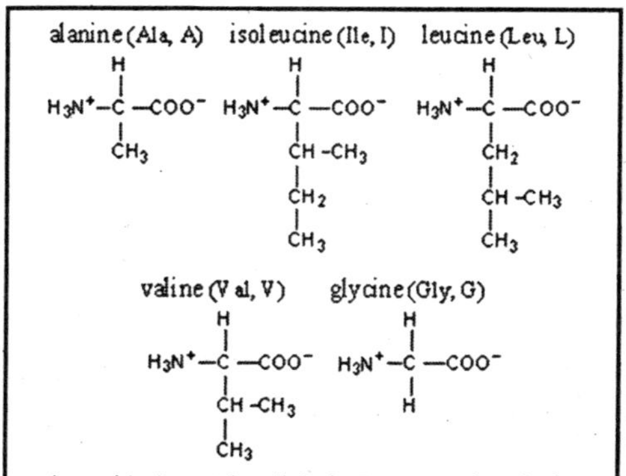

amino acids: non-polar aliphatic R-groups; plus glycine

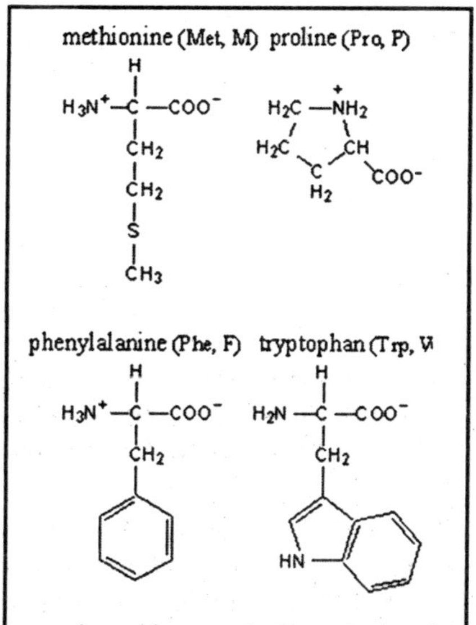

amino acids: non-polar R-groups (cont.)

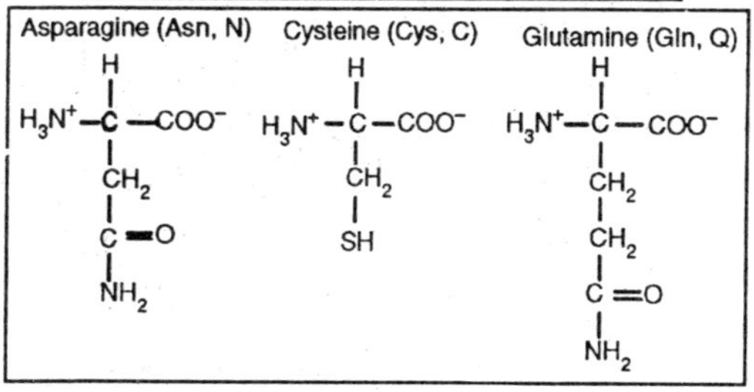

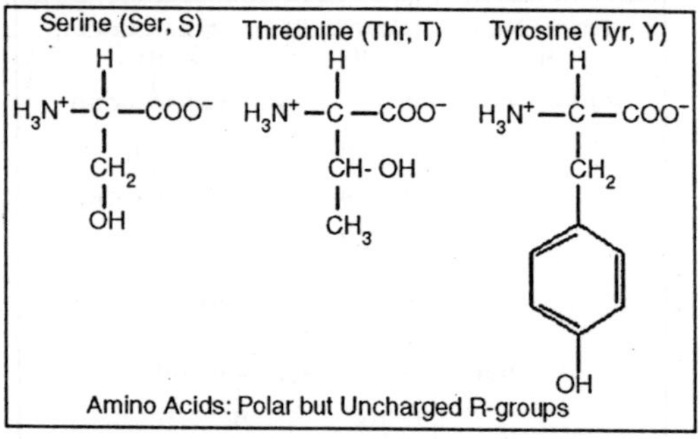

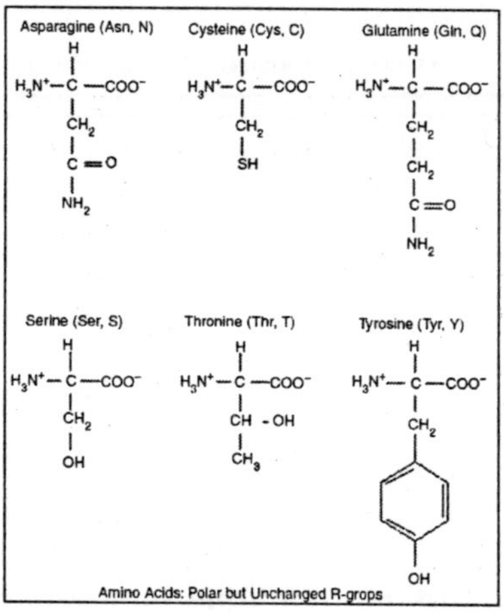

Peptide bonds link amino acid residues within proteins. The sequence of covalently linked amino acids is known as theprimary structure of a protein.The reaction by which two amino acids are joined in a peptide bond, with elimination of H_2O, is represented at right. For detail on the mechanism of peptide bond formation, see the section on protein synthesis.

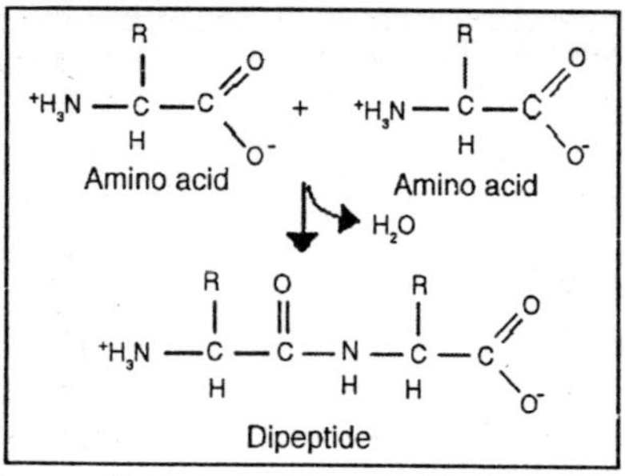

A protein is a polypeptide, a linear polymer of many amino acids, linked by peptide bonds.The peptide linkages, along with the α-carbon atoms to which R-groups are attached, form the protein backbone, with sequence NCCNCCNCCNCC.

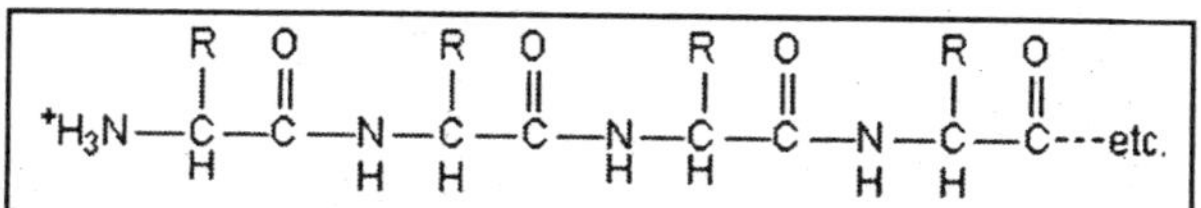

polypeptide

PROTEIN SECONDARY STRUCTURE

Segments of polypeptides often fold locally into stable structures that include α-helices and β-pleated sheets.α–Helix. In an α-helix (alpha helix), the polypeptide backbone follows a helical path. There are 3.6 amino acid residues per turn of the helix. Some protein domains assume other helical structures, but the α-helix is most common.

α-helix
(Cartoon Display)

The polypeptide chain forms a backbone structure in proteins:

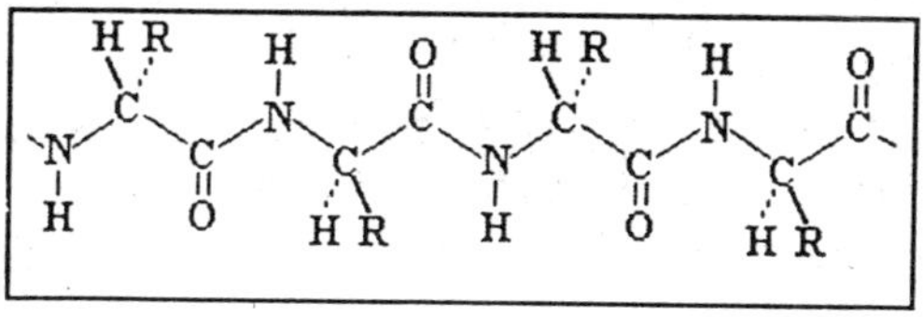

On first inspection, this structure appears to be connected entirely by single C-C or C-N bonds. It should therefore be as flexible as a simple hydrocarbon chain.Note that flexing in a covalent structure does not occur by bending bonds, and the normal tetrahedral or trigonal planar bond angles are maintained. Instead, different shapes are obtained by torsional rotationabout the axis of the bonds:

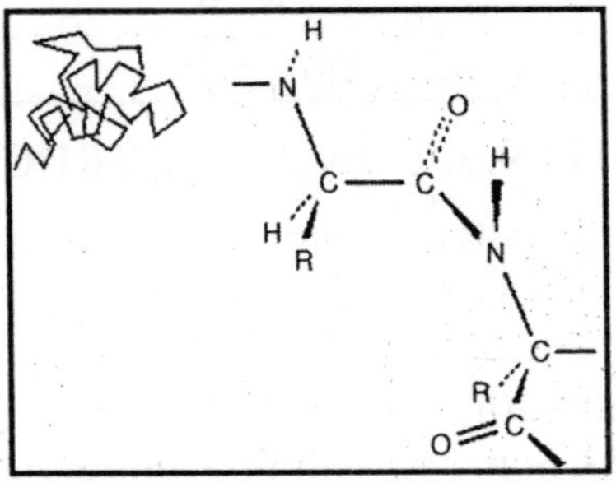

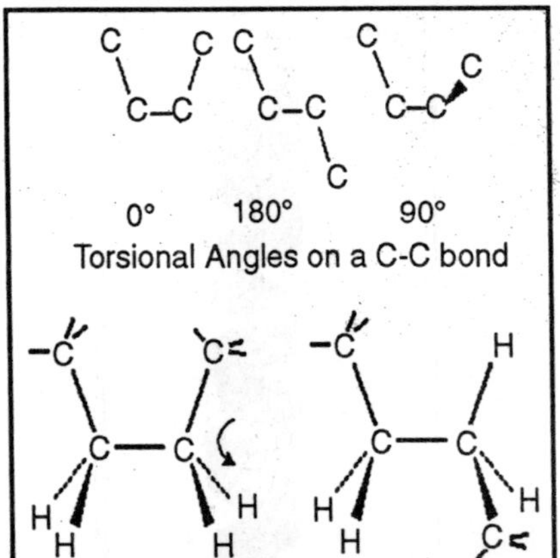

All polypeptides can adopt a form which is flexible, but random and disordered in bond orientation. This is called the denatured state of the protein. Proteins become denatured at elevated temperature or in the presence of disruptive solvents. Because there is no orderly arrangement, denatured protein is non-functional. Most proteins also have an ordered arrangement called the native state. Biochemists often work with proteins in ice baths to keep the temperature down and preserve the native state.

Orderly arrangements of the polypeptide backbone were first studied by examining fibrous proteins called keratins:

alpha-keratin - protein of hair, skin and wool;

beta-keratin or fibroin - spider and silkmoth silk.

The keratins were studied by X-ray fibre diffraction, a technique in which X-rays are reflected off a regular repetitive macromolecule array in the protein fibre, forming a characteristic pattern if the repeat spacing of the sample is comparable to X-ray wavelengths. If the X-ray wavelength is known, the size of the repeat pattern in the protein can be calculated.

For alpha-keratin, periodic repeat distances were 1.5 and 5.4 angstroms. The diffraction patterns also showed characteristic missing reflections that suggested a helical structure. For fibroin or beta keratin, periodic repeats were 3.5 and 7.0 angstroms. The structures having these repeat distances were solved and understood and its importance of knowing exact atomic radii, bond lengths and angles and used these to create exact scale models of possible structures.

However, the simple single-bonded peptide chain was so flexible that it was not initially apparent why specific structures would form. By measuring bond lengths, Pauling first deduced that the peptide bond had double bondcharacter, due to the second resonance form of the amide.

- Typical single C-N bond: 1.47 Å
- Typical double C=N bond: 1.27 Å
- Peptide bond C=O to NH: 1.33 Å

The shape of the peptide chain can be defined by the three consecutive bond torsional angles

Bond	Rotation	Torsion Angle Defined
NH to C α	Free	phi
C α to C=O	Free	psi
C=O to NH (peptide bond)	Rigid planar due to double bond character	omega

Peptide bonds are almost invariably fixed at *omega* = 180° or *trans* based on the relative alignment of C α atoms on either side of the peptide bond.

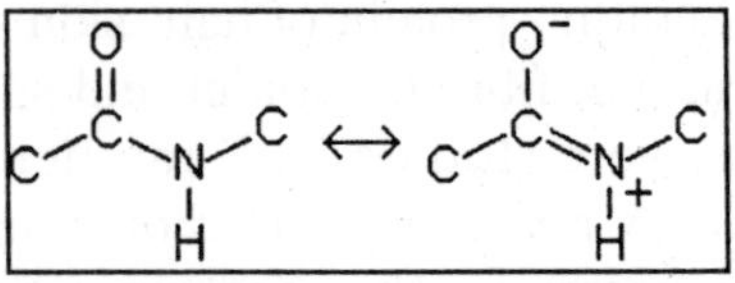

The double bond fixes the six atoms shown above into a rigid plane.In the figure below, the rigid plane of each peptide bond is indicated by the lightly shaded rectangle.C atoms are at diagonally opposite corners.

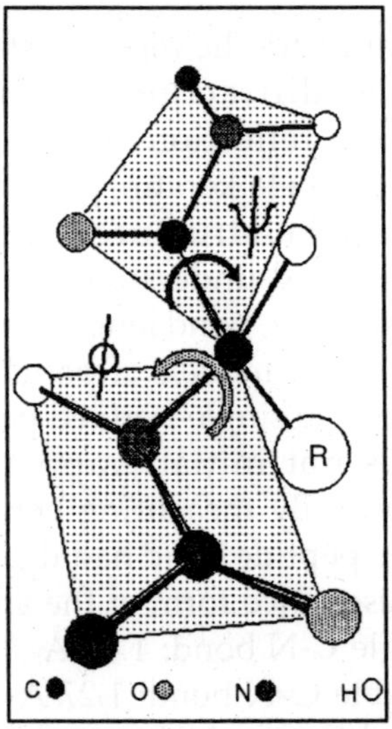

The rigidity of the peptide bond limits the number of arrangements that models could fit without distorting bonds or forcing atoms closer than van der Waals radii would allow. Without this constraint, the peptide would be free to adopt so many structures that no single consistent pattern would emerge. By reducing the degrees of freedom, a well defined set of states emerges.

Pauling found that two general patterns conformed with atomic geometry:

- An extended state for which angles phi = –135° and psi = +135°; the polypeptide chain alternatesin direction, resulting in a zig-zag structure for the peptide chain. Note the shaded circle around R; the extended strand arrangement also allows the maximum space and freedom of movement for a side chain. The repeat between identically oriented R-groups is 7.0 Å, with 3.5 Å per amino acid, matching the fibre diffraction data for beta-keratins.

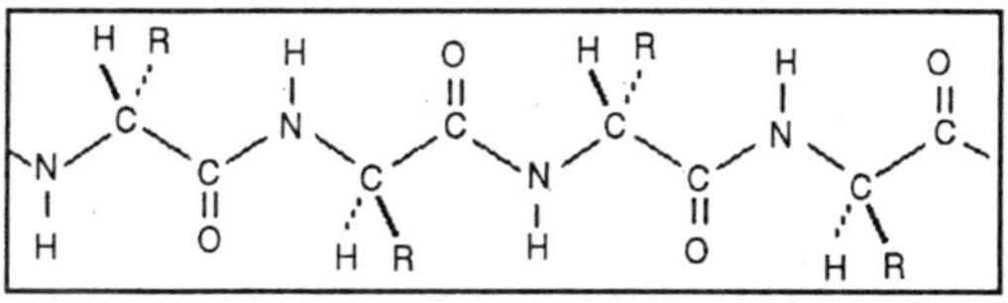

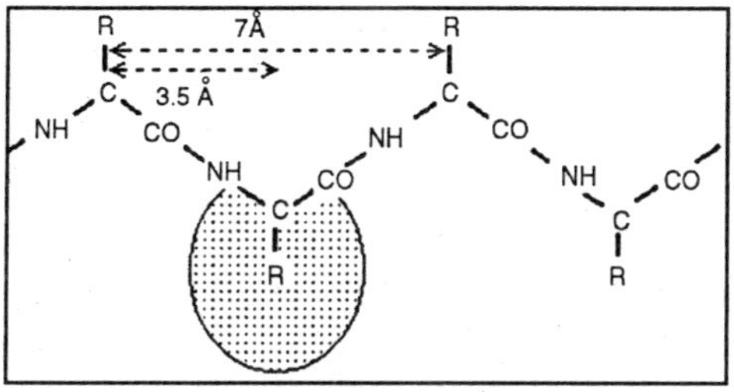

- A helical state in which the phi and psi were roughly -60°, twisting repeatedly in the same direction.The helical form models could be built with varying degrees of twist, but one model fit the atomic dimensions especially well:Alpha helix: has 3.6 amino acids per turn of the helix, which places the C=O group of amino acid #1 exactly in line with the H-N group of amino acid #5 (and C=O #2 with H-N #6).

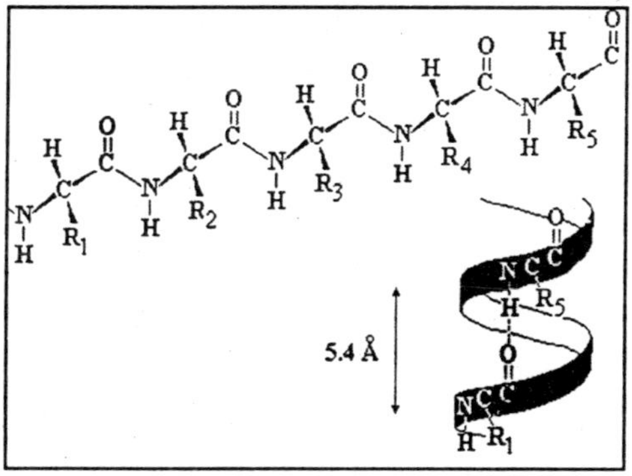

The alignment and spacing is ideal for a hydrogen bonds to form C=O:—H-N.The figure shows an alpha helix with side chains omitted for clarity. C = gray, N= blue, O = red.Hydrogen bonds are shown as dotted lines in the figure, and hydrogen bonding would make this structure especially stable.The distance separating each turn of the helix was 5.4 Å, matching periodic repeats in alpha keratin, hence the name alpha helix. Since 5.4 Å/ 3.6 is 1.5 Å, the alpha-helix has an amino acid every 1.5 Å, matching the minor periodic repeat of alpha keratin.The alpha helix was also found to be exclusively right handed; a left handed arrangement (phi and psi = +60°) has similar dimensions, but positions the amino acid side chains next to the C=O group, making the structure over-crowded.The right handed version of the helix places side chains next to the much smaller N-H, for a better fit.

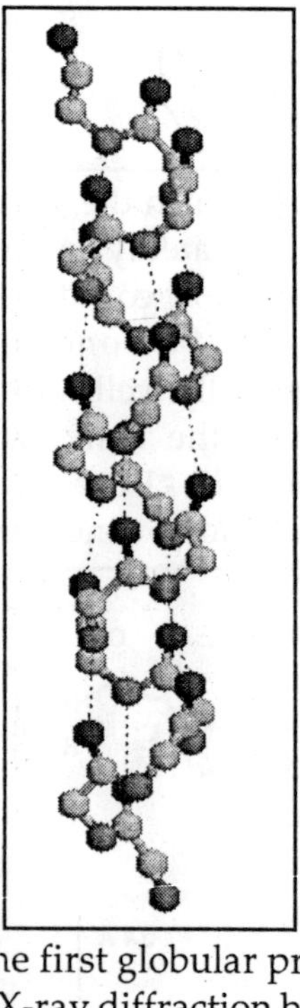

Myoglobin was the first globular protein whose structure was worked out from X-ray diffraction by protein crystals. The periodic repeats characteristic of alpha helix were recognised, and this helped determine the structure shown, in which 70% of the polypeptide is alpha-helical.

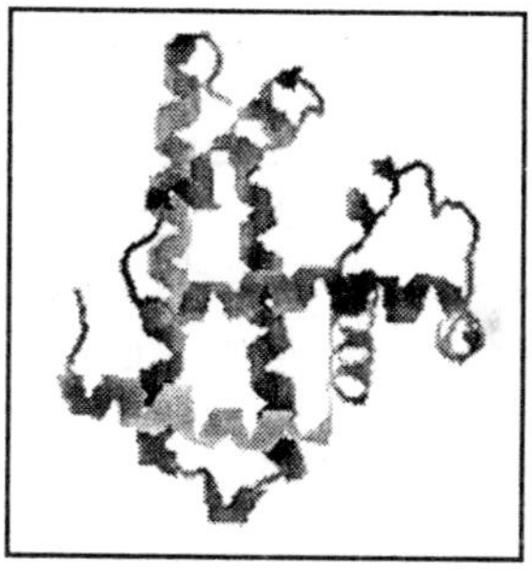

Other helices discovered by Pauling include the 2.2_7 helix, the 3_{10} helix and the pi helix. 3_{10} refers to a helix of 3 amino acids per turn, with hydrogen bonds from #1C=O to H-N #4; the hydrogen bond closes a loop of 10 atoms. 2.2_7 is a tight helix of 2.2 amino acids per turn, a 7 atom loop being closed by the H-bond #1 C=O to H-N #3. The alpha helix would be 3.6_{13} in this nomenclature; the pi helix is 4.4_{16}. Although all these helices could be modelled, experimental data was only consistent with alpha helix.

In a search of about 30000 helix segments in the protein data base:

- *Right handed alpha helix*: Almost all instances
- *Pi helix*: One instance
- *Left handed alpha helix*: One instance
- *3_{10} helix*: Only one free standing helix is known, but there are many instances of the last turn of a longer alpha helix terminating in one loop of 3_{10} helix.
- *2_7 helix*: No known instances (this is a rather strained stracture however).

In the alpha helix, the C=O---H-N bonds are almost parallel with the helix axis. The basis of hydrogen bonding is strong dipole-dipole interactions, and thus the H-bond dipoles reinforce in the helix, positive end towards the start or N-terminal end of the dipole. H-bonds are skewed relative to the helix axis in the other helices, so the reinforcement of dipoles is less effective. Pauling's extended state model matched the spacing of fibroin exactly (3.5 and 7.0 Å). In the extended state,

H-bonding NH and CO groups point out at 90º to the strand.If extended strands are lined up side by side, H-bonds bridge from strand to strand. Identical or opposed strand alignments make up parallel or antiparallel beta sheets (named for beta keratin).

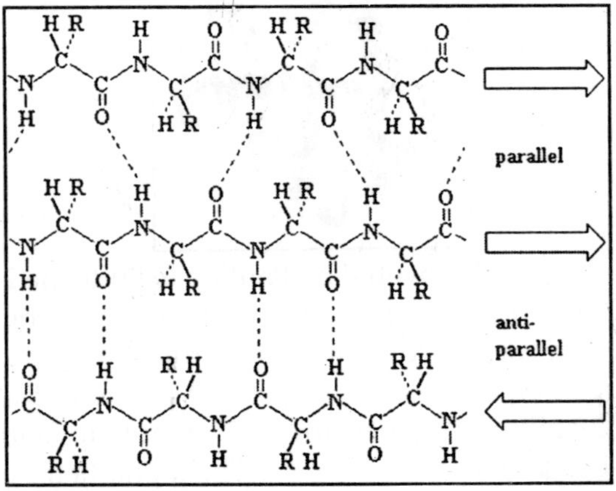

Antiparallel beta-sheet is significantly more stable due to the well aligned H-bonds. The Pauling model building studies of the early 1950's were followed up by an exhaustive computer search of *phi* and*psi* space. The analysis was based on potential energy calculations, *i.e.* through electrostatic and van der Waals interactions. Ramachandran showed that serious steric interferences occurred between C=O groups and amino acid side chain centred at phi = 120°, and serious interferences between peptide backbone CO and NH occurred at phi = psi = 0°.

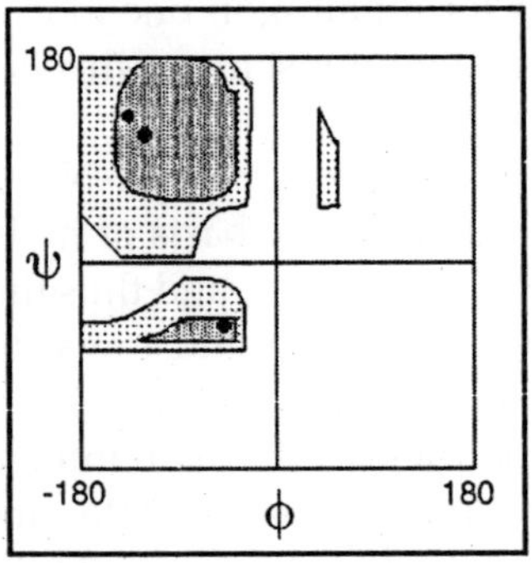

There are two regions where interactions may be described as favourable: The region of highest stability is a broad plateau centered at phi = -120° and psi = +135°. This region includes the parallel and antiparallel beta sheet.

The breadth of this region indicates that a degree of variability is allowed to the beta structure. As observed in actual proteins, beta sheets twist and flex, and beta sheets rarely actually conform exactly to the canonical extended strand. Some specialized structures such as polyglycine and polyproline are observed the edge of this zone, where phi = -60°A slightly lower but still energetically favoured region is represented by a strip running from phi = -60° to -120° and psi = -60°. The alpha helical conformation lies in this region, while 3_{10} and pi helices flank the favoured region.

In addition to the two favoured regions, two regions were calculated as allowed although not favoured. These include the mirror image of the alpha-helix strip (left handed alpha helix and the isthmus connecting the two favoured regions). The mirror image of the strongly favoured beta sheet plateau turns out to be strongly disfavoured, except by polyglycine (which has a side chain consisting only of a single H atom).

Statistical analysis of observed phi and psi angles in proteins show that most instances lie in or close to the two favoured regions (distribution shown below left), with the exception of glycine, which may adopt conformations corresponding to the mirror image of the favoured states (distribution below right).

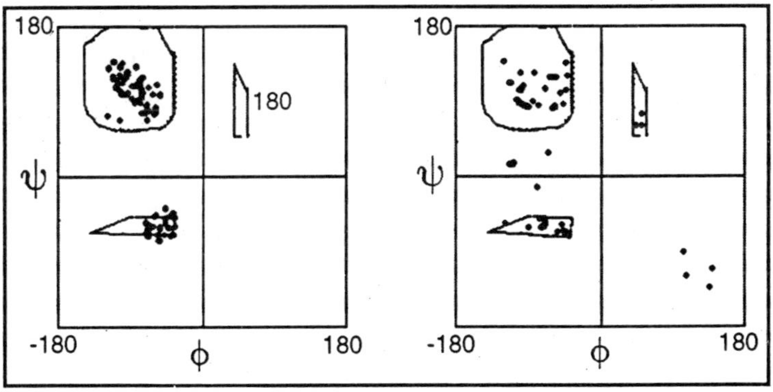

Folded proteins contain a considerable proportion of alpha helix or beta sheet: myoglobin, an alpha-helical bundle, is 70% alpha helix; other proteins may contain beta strands or a mix of alpha and beta. An alpha helix has all its amino acids repeating the same phi/psi conformation. Likewise for beta sheet, with a bit more variability due to the broad plateau in the Ramachandran plot.

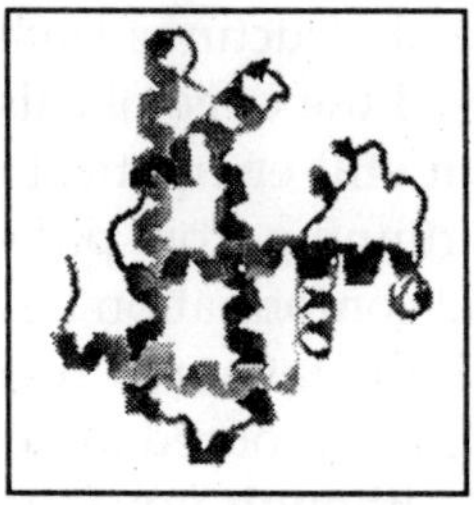

The remainder of the polypeptide exists either in the form of tight turns or loops, connecting elements that link one segment of secondary structure to the next. Tight turns consist of 4-5 amino acids in a fairly well defined structure - it's just not regular and repetitive like alpha helix or beta sheet. Loops (or random coil) are completely non-repetitive.

AMINO ACID PREFERENCES FOR DIFFERENT SECONDARY STRUCTURE

Alpha helix may be considered the default state for secondary structure. Although the potential energy is not as low as for beta sheet, H-bond formation is intra-strand, so there is an entropic advantage over beta sheet, where H-bonds must form from strand to strand, with strand segments that may be quite distant in the polypeptide sequence. The main criterion for alpha helix preference is that the amino acid side chain should cover and protect the backbone H-bonds in the core of the helix. Most amino acids do this with some key exceptions:

- *Alpha-helix preference:* Ala, Leu, Met, Phe, Glu, Gln, His, Lys,Arg

The extended structure leaves the maximum space free for the amino acid side chains: as a result, those amino acids with large bulky side chains prefer to form beta sheet structures:

- *Just plain large*: Tyr, Trp, (Phe, Met)
- *Bulky and awkward due to branched beta carbon*: Ile, Val, Thr
- *Large S atom on beta carbon*: Cys

The remaining amino acids have side chains which disrupt secondary structure, and are known as secondary structure breakers:

- Side chain H is too small to protect backbone H-bond: Gly
- Side chain linked to alpha N, has no N-H to H-bond;rigid structure due to ring restricts to phi = -60°;: Pro
- H-bonding side chains compete directly with backbone H-bonds: Asp, Asn, Ser

Clusters of breakers give rise to regions known as loops or turns which mark the boundaries of regular secondary structure, and serve to link up secondary structure segments.

There are various schemes that give the amino acids numerical weights or rankings for their preferences, and several computer programmes can predict the secondary structure from the given sequence. Examined the statistical distribution of amino acids in alpha helix, beta sheet and turns or loops, using a set of known protein structures from the protein databank.

A novel sequence can then be scanned, and the tendency of each portion of the sequence to form secondary structure is assessed. An important factor is that the final secondary structure is the consensus of all amino acids in a region. A minimum of 4 amino acids out of 6 should show alpha preference, or 3 out of 5 beta preference, or clusters of 2-3 breakers in a sequence of 4 are needed to set the secondary structure in any region, and individual misfits adopt the secondary structure of their neighbours. If you have one or two helix preference amino acids scattered in a majority of beta sheet former, the structure adopted is entirely beta sheet. More recent prediction schemes take advantage of larger data sets to examine amino acid preference for different regions in a helix or different positions in a tight turn.

In addition, sequences of homologous proteins may be compared. The rationale is that highly conserved amino acids contribute more to the three dimensional structure than unconserved, and different weightings can be introduced to the statistical analysis. The accuracy of prediction has risen from about 55% using the simple Chou-Fasman method, where the tendency is to overpredict, to about 80% using current methods. Current methods are now so complex that they are not attempted by individuals. Instead, one submits a sequence as an e-mail to certain web sites, and the results are returned in a matter of a few hours.

PROTEIN TERTIARY STRUCTURE

The tertiary structure is the complete three-dimensional structure of a polypeptide chain. Many polypeptides fold into compact, globular structures in which amino acid residues that are distant from each other in primary structure come into close proximity in the folded structure. Because of efficient packing, most water molecules are excluded from the protein's interior. It is the different interactions between the side chains of the amino acids that stabilize the tertiary structure. A major force stabilizing the tertiary structure is the hydrophobic interaction among nonpolar side chains in the core of the protein.

Background: "Principles of Protein Structure, Comparative Protein Modelling and Visualization" by N. Guex & M.C. Peitsch (GlaxoWellcome, Switzerland) is here. To obtain PDB coordinates for a protein of your interest, go to the Protein Data Bank or Molecules R Us or NCBI.

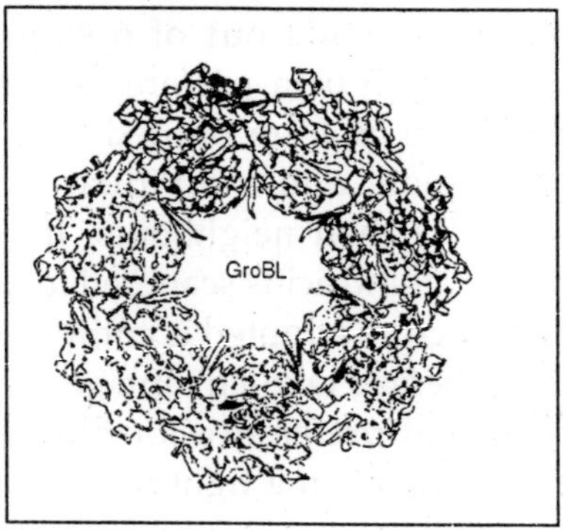

- *3D-PSSM*: (Imperial Cancer Research Fund Fold Recognition Server) A Fast, Web-based Method for protein fold recognition using 1D and 3D sequence profiles coupled with secondary structure and solvation potential information.
- *Its successor is PHYRE*: Protein Homology/analogY Recognition Engine - this is my favourite site for the prediction of the 3D structure of proteins. In each case I have used this site it has provide me with a model.
- CPHModels (Center for Biological Sequence Analysis, Technical University of Denmark) - currently consists of the following tools: Sowhat: A neural network based method to predict contacts between C-alpha atoms from the amino acid sequence. RedHom: A tool to find a subset with low sequence similarity in a database. Databases: Subsets of the Brookhaven Protein Data Bank (PDB) database with low sequence similarity produced using the RedHom tool.
- LIBELLULA (Protein Design Group. CNB-CSIC, Spain) - a neural network approach to evaluate fold recognition results. Results by e-mail. Save and view with Rasmol etc.
- LOOPP (Learning, Observing and Outputting Protein Patterns) is a fold recognition programme based on the collection of numerous signals, merging them into a single score, and generating atomic coordinates based on an alignment into a homologue template structure. The signals we are using include straightforward sequence alignment, sequence profile, threading, secondary structure and exposed surface area prediction. Excellent.
- *ESyPred3D*: This automated homology modeling programme derives benefit from a new alignment strategy using neural networks. Alignments are obtained by combining, weighting and screening the results of several multiple alignment programmes. The final three dimensional structure is built using the modeling package MODELLER..

- *AS2TS system*: Offers a variety of resources for protein structural analysis using the LGA (local-global alignment) programme to search for regions of local similarity and to evaluate the level of structural similarity between compared protein structures. To facilitate the homology-based protein structure modeling process, the AL2TS service translates given sequence–structure alignment data into the standard PDB coordinates. WHAT IF Web Interface (Centre for Molecular and Biomolecular Informatics, University of Nijmegen, Holland) offers one a large number of tools for examining PDB files.
- *Protein Peeling*: an approach for splitting a 3D protein structure into compact fragments - a method to identify small compact units (protein units (PU)) that compose protein three-dimensional structures.

InterProSurf predicts interacting amino acid residues in proteins that are most likely to interact with other proteins, given the 3D structures of subunits of a protein complex.

The prediction method is based on solvent accessible surface area of residues in the isolated subunits, a propensity scale for interface residues and a clustering algorithm to identify surface regions with residues of high interface propensities.

ProtSkin converts a protein sequence alignment in BLAST, CLUSTAL or MSF format to a property file used to map the sequence conservation onto the structure of a protein using the GRASP, MOLMOL or PyMOL. A pseudo-PDB file with the sequence conservation score in place of the temperature factor is also provided, to use with programmes such as InsightII3.

STRUCTURES DERIVED FROM NMR COORDINATES

GeNMR (GEnerate NMR structure) - generates 3D protein structures using NOE-derived distance restraints and NMR chemical shifts. Once you have a structure you may want to superimpose it on other molecules. To obtain PDB accession codes for a protein of your interest, go to the Protein Data Bank or Molecules R Us. One site is available: FATCAT (Flexible structure AlignmenT by Chaining Aligned fragment pairs allowing Twists) is an approach for flexible protein structure comparison.

It simultaneously addresses the two major goals of flexible structure alignment; optimizing the alignment and minimizing the number of rigid-body movements (twists) around pivot points (hinges) introduced in the reference structure. Dali Server - this is a network service for comparing protein structures in 3D.

You submit the coordinates of a query protein structure and Dali compares them against those in the Protein Data Bank (PDB). You receive an e-mail notification when the search has finished.

In favourable cases, comparing 3D structures may reveal biologically interesting similarities that are not detectable by comparing sequences.

Once you have a match you can use the pairwise DaliLite server to compare them. Search for Similar Protein Structures by FATCAT - either upload local PDB files or simply provide PDB codes. For close homologs go here.

PRESENTATION OF DATA

AISMIG (An Interactive Server-side Molecule Image Generator) - visualization of molecules on the Internet normally is usually done using Java Applets like jmol or webmol, Plugins like Chime or external helper applications like Rasmol which only generate low resolution pictures *i.e.* are not suitable for publication purposes. This programme produces incredible graphics.

CEP: a conformational epitope prediction server - provides a web interface to conformational epitope prediction. The algorithm, apart from predicting conformational epitopes, also predicts antigenic determinants and sequential epitopes. The epitopes are predicted using 3D structure data of protein antigens, which can be visualized graphically. The following is an example of an epitope (space-filled) mapped onto the partial surface (stick) of lysozyme:

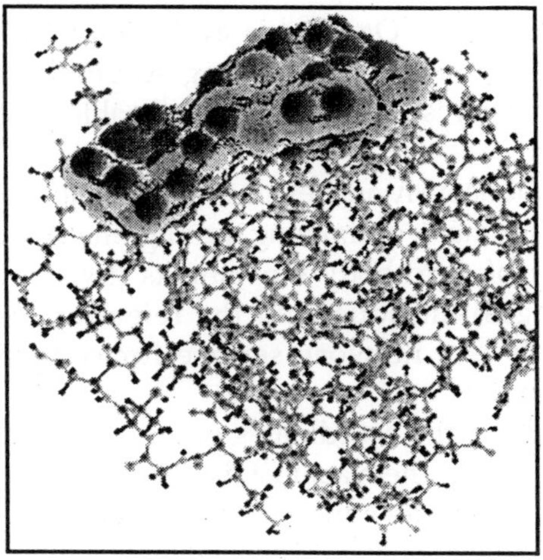

PDB2MultiGIF (Central Spectroscopy Department - Molecular Modeling, Deutsches Krebsforschungszentrum, Germany) - if you want to present your model on a webpage in a similar manner to the three pictures on the entry page use this programme. It takes the 3D structure (PDB file) and generates an animated image which can be displayed using any browser. There is considerable choice on the image size, number of frames and direction of rotation.

Protein Picture Generator - generates static or animated pictures from a protein structure file (PDB). Offers a huge number of controls on the presentation format.

MovieMaker - a web server that allows short (~10 sec), downloadable movies to be generated of protein dynamics. It accepts PDB files or PDB accession numbers as input and automatically outputs colorful animations covering a wide range of protein motions and other dynamic processes.

Users have the option of:

- Simple rotation
- Morphing between two end conformers
- Short-scale, picosecond vibrations;
- Ligand docking
- Protein oligomerization;
- Mid-scale nanosecond (ensemble) motions; and
- Protein folding/unfolding.

MovieMaker is not a molecular dynamics server and does not perform MD calculations

COMBOSA3D - Coloring Of Molecules Based On Sequence Alignment (Paul Stoddard) - accepts a group of pre-aligned sequences in FASTA format (one of the sequences must have a solved three-dimensional structure), and it uses the alignment information to highlight conserved residues on the molecule. Three-dimensional structures are shown using a Chime plugin or one may use RasMol to view and colour a molecule..

DATA MINING

- *ProFunc:* A server for predicting protein function from 3D structure - this programme takes PDB files and carries out an awesome array of analyses including scans against PDB and motif databases, determination of surface morphology and conserved residues, and potential ligand-binding sites.
- Pocket-Finder (Bioinformatics, University of Leeds, Great Britain) - is a pocket detection algorithm which works by scanning a probe radius 1.6 angstoms along all gridlines of a grid resolution 0.9 angstroms surrounding the protein. The probe also scans cubic diagonals. Grid points are defined to be part of a site when the probe is within range of protein atoms followed by free space followed by protein atoms. Since the protein is scanned in seven directions, each grid point can be defined to be part of a site up to seven times. Grid points are only retained if they are defined to be part of a site at least five times.
- *Q-SiteFinder*: Is a new method of ligand binding site prediction. It works by binding hydrophobic (CH3) probes to the protein, and finding clusters of probes with the most favourable binding energy. These clusters are placed in rank order of the likelihood of being a binding site according to the sum total binding energies for each cluster.
- *P-cats*: Prediction of CATalytic residueS in proteins -

predicts the catalytic residues in proteins from the atomic coordinates. Predict Function from Structure: ProFunc - is been developed to help identify the likely biochemical function of a protein from its three-dimensional structure using a variety of sequence- and structure-based methods.

METASITE

Scratch Protein Predictor - programmes include:

- *ACCpro*: The relative solvent accessibility of protein residues;
- *CMAPpro*: Prediction of amino acid contact maps;
- COBEpro: Prediction of continuous B-cell epitopes;
- CONpro: Predicts whether the number of contacts of each residue in a protein is above or below the average for that residue; DIpro: Prediction of disulphide bridges;
- *DISpro*: Prediction of disordered regions;
- *DOMpro*: Prediction of domains; SSpro: Prediction of protein secondary structure;
- *SVMcon*: Prediction of amino acid contact maps using Support Vector Machines; and,
- *3Dpro*: Prediction of protein tertiary structure.

BONDS

COVALENT BONDS

Covalent chemical bonds involve the sharing of a pair of valence electrons by two atoms, in contrast to the transfer of electrons in ionic bonds. Such bonds lead to stable molecules if they share electrons in such a way as to create a noble gas configuration for each atom.

Hydrogen gas forms the simplest covalent bond in the diatomic hydrogen molecule. The halogens such as chlorine also exist as diatomic gases by forming covalent bonds. The nitrogen and oxygen which makes up the bulk of the atmosphere also exhibits covalent bonding in forming diatomic molecules.

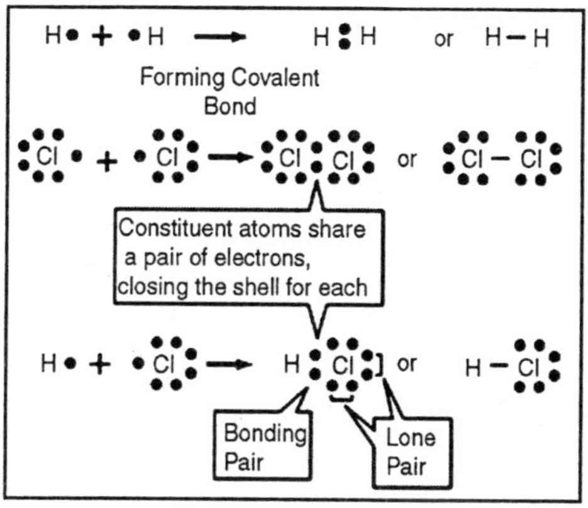

Fig. Covalent Bonding can be Visualized with the Aid of Lewis Diagrams.

IONIC BONDS

In chemical bonds, atoms can either transfer or share their valence electrons. In the extreme case where one or more atoms lose electrons and other atoms gain them in order to produce a noble gas electron configuration, the bond is called an ionic bond.Typical of ionic bonds are those in the alkali halides such as sodium chloride, NaCl.

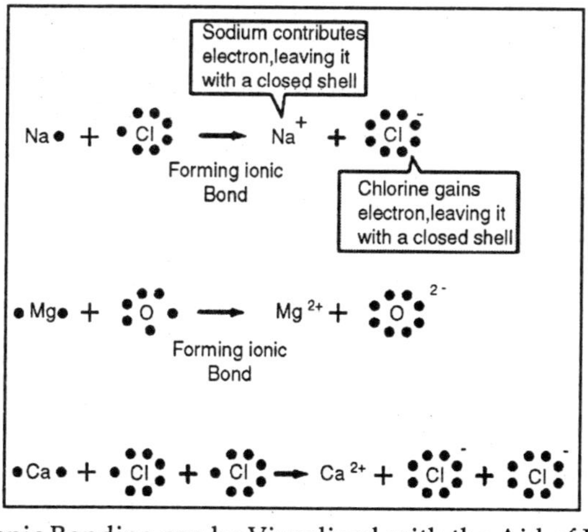

Fig. Ionic Bonding can be Visualized with the Aid of Lewis Diagrams.

HYDROGEN BONDING

Hydrogen bonding differs from other uses of the word "bond" since it is a force of attraction between a hydrogen atom in one molecule and a small atom of highelectronegativity in another molecule. That is, it is an intermolecular force, not an intramolecular force as in the common use of the word bond.

When hydrogen atoms are joined in a polar covalent bondwith a small atom of high electronegativity such as O, F or N, the partial positive charge on the hydrogen is highly concentrated because of its small size. If the hydrogen is close to another oxygen, fluorine or nitrogen in another molecule, then there is a force of attraction termed a dipole-dipole interaction. This attraction or "hydrogen bond" can have about 5% to 10% of the strength of a covalent bond.

Hydrogen bonding has a very important effect on the properties of water and ice. Hydrogen bonding is also very important in proteins and nucleic acids and therefore in life processes. The "unzipping" of DNA is a breaking of hydrogen bonds which help hold the two strands of the double helix together.

VAN DER WAALS BOND

The van der Waal bonds occur to some extent in all materials but are particularly important in plastics and polymers. These materials are made up of a long string molecules consisting of carbon atoms covalently bonded with other atoms, such as hydrogen, nitrogen, oxygen, fluorine. The covalent bonds within the molecules are very strong and rupture only under extreme conditions. The bonds between the molecules that allow sliding and rupture to occur are called van der Waal forces.

When ionic and covalent bonds are present, there is some imbalance in the electrical charge of the molecule. Take water as an example. Research has determined the hydrogen atoms are bonded to the oxygen atoms at an angle of 104.5°. This angle produces a positive polarity at the hydrogen-rich end of the molecule and a negative polarity at the other end. A result of this charge imbalance is that water molecules are attracted to each other. This is the force that holds the molecules together in a drop of water.

This same concept can be carried on to plastics, except that as molecules become larger, the van der Waal forces between molecules also increases. For example, in polyethylene the molecules are composed of hydrogen and carbon atoms in the same ratio as ethylene gas. But there are more of each type of atom in the polyethylene molecules and as the number of atoms in a molecule increases, the matter passes from a gas to a liquid and finally to a solid.

Polymers are often classified as being either a thermoplastic or a thermosetting material. Thermoplastic materials can be easily remelted for forming or recycling and thermosetting material cannot be easily remelted. In thermoplastic materials consist of long chainlike molecules. Heat can be used to break the van der Waal forces between the molecules and change the form of the material from a solid to a liquid. By contrast, thermosetting materials have a three-dimensional network of covalent bonds. These bonds cannot be easily broken by heating and, therefore, can not be remelted and formed as easily as thermoplastics.

RELATIVE BINDING ABILITY OF MOLECULAR ORBITALS

We may determine the relative importance of each orbital density in the overall binding of the nuclei in a molecule through a comparison of the forces which the various molecular orbital charge distributions exert on the nuclei. In molecular orbital theory, the total charge density is given by the sum of the orbital charge densities. Thus the total force exerted on the nuclei by the electronic charge distribution will be equal to the sum of the forces exerted by the charge density in each of the molecular orbitals. It is of interest to compare the effectiveness of each orbital charge density in binding the nuclei with some standard case in which they all exhibit the same ability. The limiting forms of the molecular orbitals for the case of the separated atoms have this desired property. In addition, the properties of the separated atoms form a useful basis for the discussion of any molecular property from the point of view of determining the changes which have been brought about by the formation of the chemical bond.

Suppose we label the two nuclei of a homonuclear diatomic molecule as A and B and consider the forces exerted on the A nucleus by the pair of electrons in a molecular orbital when the orbital has assumed its limiting form for the separated atoms. At this limit, one electron correlates with an atomic orbital on nucleus A and the other with an identical orbital on nucleus B. The discussion of the forces exerted on the nuclei by such a limiting charge distribution is similar to the discussion given previously in Section 6 for the case of two separated hydrogen atoms. The charge density in the orbital on the A nucleus will not exert a force on that nucleus since an undistorted atomic orbital is centrosymmetric with respect to its nucleus.

The charge density of the electron which correlates with the B nucleus will exert a force on the A nucleus equivalent to that obtained by concentrating the charge density to a point at the position of the B nucleus. The electron which correlates with the B nucleus will screen one of the nuclear charges of B from the A nucleus. Thus the force exerted on one of the nuclei by the pair of electrons in a molecular orbital for the limiting state of the separated atoms is equivalent to that obtained by placing one negative charge at the position of the second nucleus. Of the pair of electrons in a given homonuclear molecular orbital, only one is effective in binding either nucleus in the limit of the separated atoms.

If there are a total of N electrons in the molecule, there will be $N/2$ occupied molecular orbitals since each molecular orbital contains a pair of electrons. Therefore, a total of $N/2$ electrons will correlate with each nucleus. The molecule dissociates into two neutral atoms each with a nuclear charge $Z = N/2$. Thus the $N/2$ electrons which correlate with each nucleus will exactly cancel the nuclear charge of both nuclei; the final force on the nuclei will be zero.

The limiting force exerted on the A nucleus by the pair of electrons in a molecular orbital is $(Z_A e^2/R^2)(-1)$, that is, the force is equivalent to placing one negative charge at the position of the B nucleus. Thus we may express the total limiting force on nucleus A as the product of $(Z_A e^2/R^2)$ with the difference between the number of positive charges on the B nucleus (Z_B) and the number of electronic charges which are effective in exerting a force on *the A nucleus,* $(N/2)$:

$$F_A = \frac{Z_A e^2}{R^2}(Z_B - (N/2))$$

The quantity $N/2$ is the *charge equivalent* of the electronic force, the number of charges which when placed at the position of one nucleus exerts the same force on the second nucleus as does the actual charge distribution. The zero force between the separated atoms may be viewed as a result of each electron screening one nuclear charge on one nucleus from the nuclear charge of the other atom.

As the atoms approach one another to form a chemical bond, the atomic distributions on each atom become increasingly distorted and charge density is transferred to the binding region between the nuclei. There is a net force of attraction on the nuclei. We may again express the electronic force on the A nucleus in terms of its charge equivalent by multiplying the electronic force of attraction by $R^2/Z_A e^2$. Because of the distortion of the atomic orbital densities and the formation of molecular orbitals concentrating charge density in the region between the nuclei, the charge density of more than just one electron in each molecular orbital is effective in binding the nuclei. Thus at intermediate R values the charge equivalent of the electronic force exceeds its limiting value of $N/2$ required to screen the nuclear charge and the result is a force of attraction drawing the nuclei together.

When the distance between the nuclei is further decreased to its equilibrium value the force on the nuclei is again equal to zero. At this point, as when R equals infinity, the charge equivalent of the electronic force equals the nuclear charge. However, the state of electrostatic equilibrium in the molecule does not correspond to the charge density in each molecular orbital effectively screening one nuclear charge as it did in the separated atoms. Instead the charge equivalent of the density in each molecular orbital may be less than, equal to, or greater than the limiting value of unity observed for the separated atoms.

An orbital which concentrates charge density in the binding region will exert a force on the nuclei with a charge equivalent greater than unity. Such an orbital is called *binding* as it does more than simply screen one unit of positive charge on each nucleus.

The charge equivalent of an orbital which concentrates density in the antibinding regions will be less than the separated atom value of unity. Such an orbital is termed *antibinding* as the charge density does not screen one unit of positive charge on each nucleus. When the charge equivalent of the force equals unity, it implies that the orbital charge density plays the same role in the molecule as in the separated atoms, that of screening one nuclear charge on B from nucleus A. An orbital with this property is termed *nonbinding*.

Thus, by comparing the charge equivalent of the force exerted by the density in each molecular orbital with its separated atom value of unity, we may classify the orbitals as binding, antibinding or nonbinding:

Binding orbital	—	charge equivalent > unity
Nonbinding orbital	—	charge equivalent ~ unity
Antibinding orbital	—	charge equivalent < unity

A = Surface area available

d = Thickness of membrane

C1-C2 =concentration gradient

Note: For a given membrane A and will be constant

Example Passive Process Table. Charge Equivalents of the Orbital Forces in Homonuclear Diatomic Molecules.

Molecule	$1\sigma_g$	$1\sigma_u$	$2\sigma_g$	$2\sigma_u$	$1p_u$	$3\sigma_g$	$1p_g$	Sum = Z_B	R_e (au)
He_2	1.78	-0.42						1.36*	(1.750)*
Li_2	0.70	0.68	1.62					3.00	5.051
Be_2	1.05	1.08	2.00	-0.40				3.68*	(3.500)*
B_2	0.98	0.98	2.32	-0.48	1.20			5.00	3.005
C_2	0.97	0.95	2.25	-0.43	1.13†			6.00	2.348
N_2	1.15	1.08	2.67	-0.47	1.21†	0.15		7.00	2.068
O_2	1.23	1.14	2.94	-0.52	1.30†	0.18	0.43	8.00	2.282
F_2	1.24	1.12	2.45	-0.16	1.24†	0.52	0.67†	9.00	2.680

*He_2 and Be_2 form only weakly bound van der Waals molecules for relatively large internuclear seperations. The values of$R$ quoted for these molecules are the internuclear distances used in the calculation of the charge equivalents listed in the table.

All of the values are quoted for double occupation of the orbitals for comparative purposes. The values marked by †are to be doubled to obtain the total electronic force as they refer to filled π orbitals.

The charge equivalents of the orbital forces for some homonuclear diatomic molecules are given in Table above. Except for He_2 and Be_2 the sum of the charge equivalents equals the nuclear charge in each case as required for electrostatic equilibrium and the formation of a stable molecule. The charge equivalents of the orbital forces provide a quantitative measure of the role each orbital density plays in the binding of the nuclei in the molecule.

The $1\sigma_g$ orbital in He_2 is binding. Of the two electronic charges in the $1\sigma_g$ orbital, 1.78 of them are effective in binding the nuclei when $R = 1.75\ a_0$ as opposed to the one electronic charge which exerts a force when $R = \infty$. The $1\sigma_u$ charge density, however, is strongly antibinding. The transfer of charge density to the antibinding regions in the formation of the $1\sigma_u$ orbital in He_2 is so great that the charge equivalent is negative in sign. The antibinding nature of this orbital is very evident in the form of its charge distribution. Not only does the charge density in this orbital no longer screen a positive charge on one nucleus from the other, it actually exerts a repulsive force on the nuclei, one which pulls the nuclei further apart from one another. The total eletronic force exerted on a nucleus in He_2 at $R = 1.75\ a_o$ is equivalent to placing (1.78 - 0.42) = 1.36 negative charges at the position of the second nucleus. Since the nuclear charge on helium is 2.00, a total of (2.00 - 1.36) = 0.64 positive charges on the second nucleus are left unscreened by the charge density. The net force on the nuclei is thus a repulsive one.

The $1\sigma_g$ and $1\sigma_u$ molecular orbitals are inner shell orbitals in the remaining molecules, Li_2 to F_2. An idealized inner shell molecular orbital has a charge equivalent of unity, the same as the separated atom value. Each electron should be localized in an atomic-like distribution and screen one nuclear charge. This is illustrated by the $1\sigma_g$ and $1\sigma_u$ charge density maps for the O_2 molecule. The charge equivalents of the $1\sigma_g$ and $1\sigma_u$ orbital densities for Li_2 are significantly less than unity. While these orbitals are not as contracted around the nuclei in Li_2 as they are in O_2 (the nuclear charge for lithium is three compared to eight for oxygen), they are still atomic-like with no effective overlap between the two centres. The charge equivalents are less than the screening value of unity because each of the atomic-like distributions is polarized into the antibinding region and exerts an antibinding force on the nucleus on which it is centred. The charge equivalents for the $1\sigma_g$ and $1\sigma_u$ density distributions in the remaining molecules are close to unity indicating that they are essentially nonbinding inner shell orbitals. The slight binding character of the $1\sigma_g$ charge density in O_2 and F_2 is the result of small inward polarizations of the atomic-like distributions.

The $2\sigma_g$ molecular charge density is binding in every molecule. A comparison of the charge equivalents shows that the $2\sigma_g$ charge density is the most binding of all the molecular orbitals in this series of molecules. The charge equivalent of the force exerted by the $2\sigma_g$ density in O_2 is almost three times greater than it is for the separated oxygen atoms. This is a result of the large amount of charge density accumulated in the binding region by this orbital.

The $2\sigma_u$ orbital is uniformly strongly antibinding. The extreme concentration of charge density in the antibinding regions observed for the $2\sigma_u$ orbital is typified by the

$2\sigma_u$ density plot for O_2. It is obvious that the density in this orbital, as that in the $1\sigma_u$ orbital of He_2 will pull the nuclei away from one another rather than bind them together. Notice that Be_2 is analogous to He_2 except that the $2\sigma_g$ and $2\sigma_u$ orbitals rather than the $1\sigma_g$and $1\sigma_u$ orbital densities are involved. In Be_2 the $1\sigma_g$ and $1\sigma_u$ densities are nonbinding and together simply screen two nuclear charges on each atom. The $2\sigma_g$ density exerts a binding force equivalent to one electronic charge in excess of the simple screening effect. The $2\sigma_u$ orbital density, however, leaves a single nuclear charge unscreened which cancels the net attractive force of the $2\sigma_g$density and in addition exerts an antibinding force equivalent to increasing the nuclear charge by 0.40 units. The beryllium molecule is therefore unstable at this value of R.

The $1\pi_u$ orbital density is binding in each case, but only weakly so. The charge density of a π_u molecular orbital is concentrated around the internuclear axis rather than along it as in a σ_g molecular orbital. Consequently the $1\pi_u$ density distributions exert only weak binding forces on the nuclei. In fact, the inner shell $1\sigma_g$ charge density in F_2 exerts as large a binding force on the nuclei as does a pair of electrons in the$1\pi_u$ orbital.

The charge equivalent of the 3 σ_g orbital density is less than unity in the three cases where it is occupied. Thus it is an antibinding orbital even though it is of σ_g symmetry. The charge density contours for this orbital in O_2 show that charge density is accumulated in the region between the nuclei as expected for an orbital of σ_g symmetry. However, the 3 σ_g orbital correlates with a $2p$ σ atomic orbital on each nucleus. The strong participation of the $2p$ σorbitals in the molecular orbital is evidenced by the node at each nucleus and by the concentration of charge density on both sides of each nucleus. The concentration of charge in the antibinding regions nullifies the binding effect arising from the accumulation of charge density in the region between the nuclei. The net result is an attractive force considerably less than that required to screen one positive charge on each nucleus. The $1\pi_g$ orbital density is only weakly antibinding just as the $1\pi_u$ density is only weakly binding. The formation of the σ_g orbital results in the removal of charge density from the binding region, not from along the internuclear axis but instead from regions around the axis. Notice that unlike the $2\sigma_u$ orbital densities, the $1\pi_g$ charge density is antibinding only in the sense that it does not screen its share of nuclear charge, not because it exerts a force which draws the nuclei apart.

QUATERNARY POTEIN STRUUCTURE

The 2D projections of the computed interface aid visualization of complicated interfacial geometries in 3D. Molecular properties, including hydrophobicity and electrostatic potential, can be projected onto the interface. MolSurfer can thereby aid exploration of molecular complementarity, identification of binding "hot spots" and prediction of the effects of mutations. MolSurfer can also facilitate the location of cavities at macromolecular interfaces.

PQS PROTEIN QUATERNARY STRUCTURE (EMBL-EBI)

- *Inter PreTS prediction*: Combines three-dimensional information of protein complexes, from their database of interacting domains, with an empirical scoring system to assess the fit of any potential protein interacting pair with a known three-dimensional structure.
- *String*: A Search Tool for the Retrieval of Interacting Genes/Proteins - Associations are derived from high-throughput experimental data, mining databases and literature, and from predictions based on genomic context analysis. STRING integrates and ranks these associations by benchmarking them against a common reference set, and presents evidence in a consistent and intuitive web interface. Importantly, the associations are extended beyond the organism in which they were originally described, by automatic transfer to orthologous protein pairs in other organisms, where applicable.meta-PPISP is built on three individual web servers: cons-PPISP, PINUP,

and Promate. This is a meta web server for protein-protein interaction site prediction.

MOLECULAR STUCTURE OF PROTEINS

Virtually every aspect of cellular activity is dependent on a protein or a set of proteins. Proteins are true agents of biological function and are involved in a great variety of cellular functions including the catalysis of reactions (enzymes), regulation (hormones), transportation, storage, structural function, and protection (immunoglobulin). Each protein is coded by one gene. Proteins are the expression of the genetic information contained in the DNA molecule.

AMINO ACIDS: THE BUILDING BLOCKS OF PROTEINS

From a chemical point of view, proteins are linear (unbranched) polymers of amino acids. Each amino acid becomes attached to another when the amino end (NH2) of one amino acid joins to the carboxyl end (COOH) of another. Amino acids possess an R chain (R for radical) and a hydrogen atom and all are attached to a central carbon atom. It is in the R chain where the variety of amino acids are formed. Differences in the structure of the R chains (sometimes known as the amino acid side chain) result in different amino acid properties.

IMPORTANCE OF PRIMARY STRUCTURE

All the structural levels of a protein (secondary, tertiary and quaternary) are ultimately determined by the primary structure because all the information needed for the molecule to achieve its conformation is imprinted within the primary structure. It is this polypeptide chain (and its components) that determines where the proteins bend, fold, and where they can link to another polymer chain.

7

Hormones

PLANT-HORMONES

The plant hormones was set up initially at institute of the BBSRC. Long Ashton will be closing in March 2003 so the Plant-Hormones site has been moved to a new commercial service. The general information on plant hormones and links to the plant-hormones has been set up. This list has been set up as a freely accessible electronic communication forum for use by plant-hormone scientists. The list is intended to promote communication between professionals in the plant-hormone field. Discussion topics should include gibberellins, auxins, cytokinins,abscisic acid, etc. All aspects of research may be discussed; synthesis extraction, purification, analysis, enzymology, molecular biology, hormone action etc. Requests for help on particular aspects may also be posted to the list as well as requests for standards, spectra, identification of spectra etc. The list may also be used for announcement of meetings, calls for papers and posters, visits of scientists to a particular area.

ABSCISIC ACID

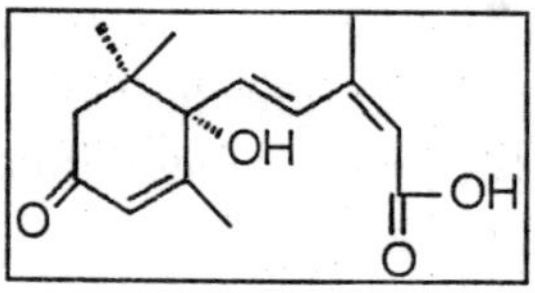

Nature of Abscisic Acid

Abscisic acid is a single compound unlike the auxins, gibberellins, and cytokinins. It was called"abscisin II" originally because it was thought to play a major role in abscission of fruits. At about the same time another group was calling it"dormin" because they thought it had a major role in bud dormancy. The name abscisic acid (ABA) was coined by a compromise between the two groups. Though ABA generally is thought to play mostly inhibitory roles, it has many promoting functions as well.

History of Abscisic Acid

In 1963, abscisic acid was first identified and characterized by Frederick Addicott and his associates. They were studying compounds responsible for the abscission of fruits (cotton). Two compounds were isolated and called abscisin I and abscisin II. Abscisin II is presently called abscisic acid (ABA). Two other groups at about the same time discovered the same compound. One group headed by Philip Wareing was studying bud dormancy in woody plants. The other group led by Van Steveninck was studying abscission of flowers and fruits from lupine. Plant physiologists agreed to call the compound abscisic acid.

Biosynthesis and Metabolism

ABA is a naturally occurring compound in plants. It is a sesquiterpenoid (15-carbon) which is partially produced via the mevalonic pathway in chloroplasts and other plastids. Because it is sythesized partially in the chloroplasts, it makes sense that biosynthesis primarily occurs in the leaves. The production of ABA is accentuated by stresses such as water loss and freezing temperatures. It is believed that biosynthesis occurs indirectly through the production of carotenoids. Carotenoids are pigments produced by the chloroplast which have 40 carbons. Breakdown of these carotenoids occurs by the following mechanism:

Violaxanthin is a carotenoid which has forty carbons. It is isomerized and then split via an isomerase reaction followed by an oxidation reaction. One molecule of xanthonin is produced from one molecule of violaxanthonin and it is uncertain what happens to the remaining biproduct. The one molecule of xanthonin produced is unstable and spontaneously changed to ABA aldehyde. Further oxidation results in ABA.

Activation of the molecule can occur by two methods. In the first method, an ABA-glucose ester can form by attachment of glucose to ABA. In the second method, oxidation of ABA can occur to form phaseic acid and dihyhdrophaseic acid. The transport of ABA can occur in both xylem and phloem tissues. It can also be translocated through paranchyma cells. The movement of abscisic acid in plants does not exhibit polarity like auxins. ABA is capable of moving both up and down the stem.

Functions of Abscisic Acid

The following are some of the phyysiological responses known to be associated with abscisic acid:

- Stimulates the closure of stomata (water stress brings about an increase in ABA synthesis).
- Inhibits shoot growth but will not have as much affect on roots or may even promote growth of roots.
- Induces seeds to synthesize storage proteins.
- Inhibits the affect of gibberellins on stimulating de novo synthesis of a-amylase.
- Has some effect on induction and maintanance of dormancy.

Induces gene transcription especially for proteinase inhibitors in response to wounding which may explain an apparent role in pathogen Defence.

AUXINS

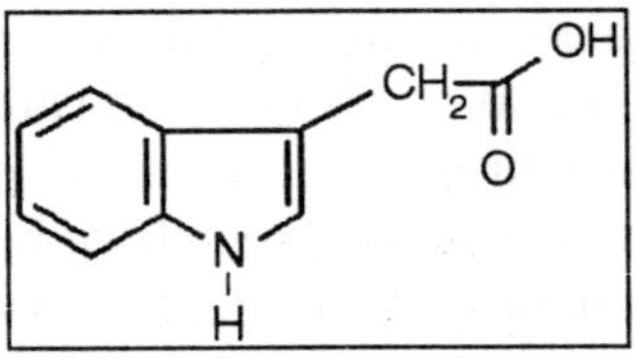

Nature of Auxins

The term auxin is derived from the Greek word auxein

which means to grow. Compounds are generally considered auxins if they can be characterized by their ability to induce cell elongation in stems and otherwise resemble indoleacetic acid (the first auxin isolated) in physiological activity. Auxins usually affect other processes in addition to cell elongation of stem cells but this characteristic is considered critical of all auxins and thus"helps" define the hormone.

History of Auxins and Pioneering Experiments

Auxins were the first plant hormones discovered. Charles Darwin was among the first scientists to dabble in plant hormone research. In his book"The Power of Movement in Plants" presented in 1880, he first describes the effects of light on movement of canary grass coleoptiles. The coleoptile is a specialized leaf originating from the first node which sheaths the epicotyl in the plants seedling stage protecting it until it emerges from the ground.

When unidirectional light shines on the coleoptile, it bends in the direction of the light. If the tip of the coleoptile was covered with aluminum foil, no bending would occur towards the unidirectional light. However if the tip of the coleoptile was left uncovered but the portion just below the tip was covered, exposure to unidirectional light resulted in curvature towards the light. Darwin's experiment suggested that the tip of the coleoptile was the tissue responsible for perceiving the light and producing some signal which was transported to the lower part of the coleoptile where the physiological response of bending occurred.

He then cut off the tip of the coleoptile and exposed the rest of the coleoptile to unidirectional light to see if curving occurred. Curvature did not occur confirming the results of his first experiment.

It was in 1885 that Salkowski discovered indole-3-acetic acid (IAA) in fermentation media. The isolation of the same product from plant tissues would not be found in plant tissues for almost 50 years. IAA is the major auxin involved in many of the physiological processes in plants. In 1907, Fitting studied the effect of making incisions on either the light or dark side of the plant. His results were aimed at understanding if translocation of the signal occurred on a particular side of the plant but his results were inconclusive because the signal was capable of crossing or going around the incision.

In 1913, Boysen-Jensen modified Fritting's experiment by inserting pieces of mica to block the transport of the signal and showed that transport of auxin towards the base occurs on the dark side of the plant as opposed to the side exposed to the unidirectional light. In 1918, Paal confirmed Boysen-Jensen's results by cutting off coleoptile tips in the dark, exposing only the tips to the light, replacing the coleoptile tips on the plant but off Centreed to one side or the other.

Results showed that whichever side was exposed to the coleoptile, curvature occurred towards the other side. Soding was the next scientist to extend auxin research by extending on Paal's idea. He showed that if tips were cut off there was a reduction in growth but if they were cut off and then replaced growth continued to occur. In 1926, a graduate student from Holland by the name of Fritz Went published a report describing how he isolated a plant growth substance by placing agar blocks under coleoptile tips for a period of time then removing them and placing them on decapitated Avena stems. After placement of the agar, the stems resumed growth

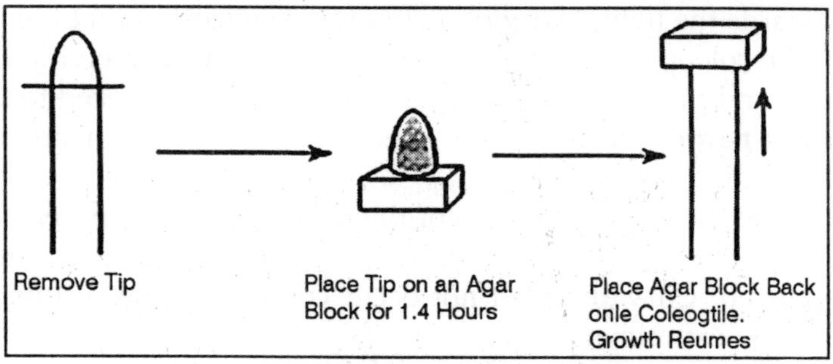

In 1928, Went developed a method of quantifying this plant growth substance. His results suggested that the curvatures of stems were proportional to the amount of growth substance in the agar. This test was called the avena curvature test.(see below) Much of our current knowledge of auxin was obtained from its applications. Went's work had a great influence in stimulating plant growth substance research. He is often credited with dubbing the term auxin but it was actually Kogl and Haagen-Smit who purified the compound auxentriolic acid (auxin A) from human urine in 1931.

Later Kogl isolated other compounds from urine which were similar in structure and function to auxin A, one of which was indole-3 acetic acid (IAA) initially discovered by Salkowski in 1985. A committee of plant physiologists was set up to characterize the group auxins. The term comes from the Greek auxein meaning"to grow." Compounds are generally considered auxins if they are synthesized by the plant and are substances which share similar activity to IAA.

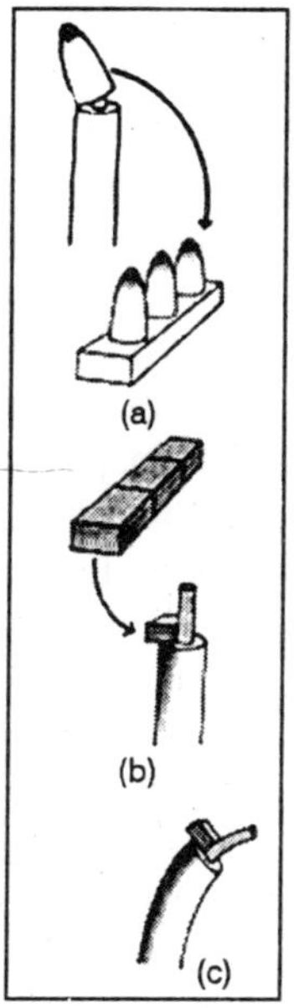

BIOSYNTHESIS AND METABOLISM OF AUXIN

IAA is chemically similar to the amino acid tryptophan which is generally accepted to be the molecule from which IAA is derived. Three mechanisms have been suggested to explain this conversion: Tryptophan is converted to indolepyruvic acid through a transamination reaction. Indolepyruvic acid is then converted to indoleacetaldehyde by a decarboxylation reaction.

The final step involves oxidation of indoleace-taldehyde resulting in indoleacetic acid. Tryptophan undergoes decarboxylation resulting in tryptamine. Tryptamine is then oxidized and deaminated to produce indoleacetaldehyde. This molecule is further oxidized to produce indoleacetic acid. As recently as 1991, this 3rd mechanism has evolved. IAA can be produced via a tryptophan-independent mechanism. This mechanism is poorly understood, but has been proven using trp(-) mutants. Other experiments have shown that, in some plants, this mechanism is actually the preferred mechanism of IAA biosynthesis.

The enzymes responsible for the biosynthesis of IAA are most active in young tissues such as shoot apical meristems and growing leaves and fruits. The same tissues are the locations where the highest concentrations of IAA are found. One way plants can control the amount of IAA present in tissues at a particular time is by controlling the biosynthesis of the hormone. Another control mechanism involves the production of conjugates which are, in simple terms, molecules which resemble the hormone but are inactive. The formation of conjugates may be a mechanism of storing and transporting the active hormone. Conjugates can be formed from IAA via hydrolase enzymes. Conjugates can be rapidly activated by environmental stimuli signaling a quick hormonal response. Degradation of auxin is the final method of controlling auxin levels.

This process also has two proposed mechanisms outlined below: The oxidation of IAA by oxygen resulting in the loss of the carboxyl group and 3-methyleneoxindole as the major breakdown product. IAA oxidase is the enzyme which catalyzes this activity. Conjugates of IAA and synthetic auxins such as 2,4-D can not be destroyed by this activity. C-2 of the heterocyclic ring may be oxidized resulting in oxindole-3-acetic acid. C-3 may be oxidized in addition to C-2 resulting in dioxindole-3-acetic acid. The mechanisms by which biosynthesis and degradation of auxin molecules occur are important to future agricultural applications. Information regarding auxin metabolism will most likely lead to genetic and chemical manipulation of endogenous hormone levels resulting in desirable growth and differentiation of important crop species. Ultimately, the possibility exists to regulate plant growth without the use of hazardous herbicides and fertilizers.

Functions of Auxin

The following are some of the responses that auxin is known to cause:

- Stimulates cell elongation
- Stimulates cell division in the cambium and, in combination with cytokinins in tissue culture
- Stimulates differentiation of phloem and xylem.
- Stimulates root initiation on stem cuttings and lateral root development in tissue culture.
- Mediates the tropistic response of bending in response to gravity and light.
- The auxin supply from the apical bud suppresses growth of lateral buds.
- Delays leaf senescence.
- Can inhibit or promote (via ethylene stimulation) leaf and fruit abscission.
- Can induce fruit setting and growth in some plants.
- Involved in assimilate movement towards auxin. possibly by an effect on phloem transport.
- Delays fruit ripening.
- Promotes flowering in Bromeliads.
- Stimulates growth of flower parts.
- Promotes (via ethylene production) femaleness in dioecious flowers.
- Stimulates the production of ethylene at high concentrations.

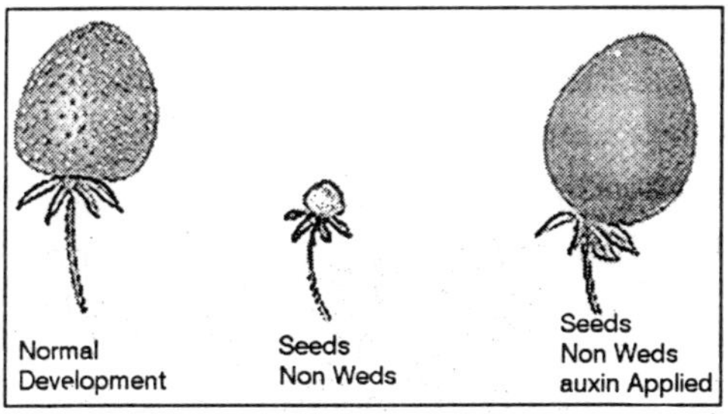

Above describes the effect of auxin on strawberry development. The achenes produce auxin. When removed the strawberry does not develop.

CYTOKININS

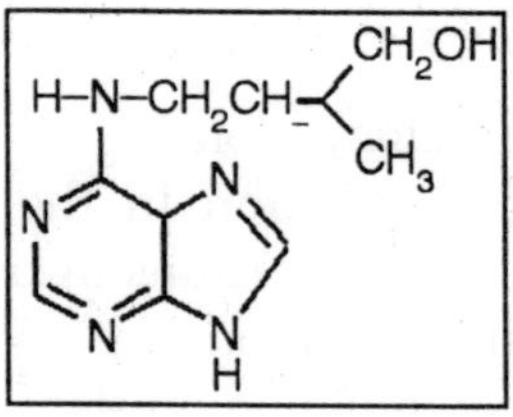

NATURE OF CYTOKININS

Cytokinins are compounds with a structure resembling adenine which promote cell division and have other similar functions to kinetin. Kinetin was the first cytokinin discovered and so named because of the compounds ability to promote cytokinesis (cell division). Though it is a natural compound, It is not made in plants, and is therefore usually considered a"synthetic" cytokinin (meaning that the hormone is synthesized somewhere other than in a plant). The most common form of naturally occurring cytokinin in plants today is called zeatin which was isolated from corn.

Cytokinins have been found in almost all higher plants as well as mosses, fungi, bacteria, and also in tRNA of many prokaryotes and eukaryotes. Today there are more than 200 natural and synthetic cytokinins combined. Cytokinin concentrations are highest in meristematic regions and areas of continuous growth potential such as roots, young leaves, developing fruits, and seeds.

HISTORY OF CYTOKININS

Gottlieb Haberlandt discovered that a compound found in phloem had the ability to stimulate cell division. Johannes van Overbeek discovered that the milky endosperm from coconut also had this ability. He also showed that various other plant species had compounds which stimulated cell division. In 1954, Jablonski and Skoog extended the work of Haberlandt showing that vascular tissues contained compounds which promote cell division. The first cytokinin was isolated from herring sperm by Miller and his associates. This compound was named kinetin because of its ability to promote cytokinesis. Hall and deRopp reported that kinetin could be formed from DNA degradation products. The first naturally occurring cytokinin was isolated from corn by Miller.

It was later called zeatin. Almost simultaneous with Miller Letham published a report on zeatin as a factor inducing cell division and later described its chemical properties. It is Miller and Letham that are credited with the simultaneous discovery of zeatin. Since that time, many more naturally occurring cytokinins have been isolated and the compound is ubiquitous to all plant species in one form or another.

BIOSYNTHESIS AND METABOLISM OF CYTOKININS

Cytokinin is generally found in higher concentrations in meristematic regions and growing tissues. They are believed to be synthesized in the roots and translocated via the xylem to shoots. Cytokinin biosynthesis happens through the biochemical modification of adenine. The process by which they are synthesized is as follows:

A product of the mevalonate pathway called isopentyl pyrophosphate is isomerized. This isomer can then react with adenosine monophosphate with the aid of an enzyme called isopentenyl AMP synthase. The result is isopentenyl adenosine-5'-phosphate (isopentenyl AMP). This product can then be converted to isopentenyl adenosine by removal of the phosphate by a phosphatase and further converted to isopentenyl adenine by removal of the ribose group. Isopentenyl adenine can be converted to the three major forms of naturally occurring cytokinins. Other pathways or slight alterations of this one probably lead to the other forms. Degradation of cytokinins occurs largely due to the enzyme cytokinin oxidase. This enzyme removes the side chain and releases adenine. Derivitives can also be made but the pathways are more complex and poorly understood.

CYTOKININ FUNCTIONS

A list of some of the known physiological effects caused by cytokinins are listed below. The response will vary depending on the type of cytokinin and plant species.

- Stimulates cell division.
- Stimulates morphogenesis (shoot initiation/bud formation) in tissue culture.
- Stimulates the growth of lateral buds-release of apical dominance.
- Stimulates leaf expansion resulting from cell enlargement.
- May enhance stomatal opening in some species.
- Promotes the conversion of etioplasts into chloroplasts via stimulation of chlorophyll synthesis.

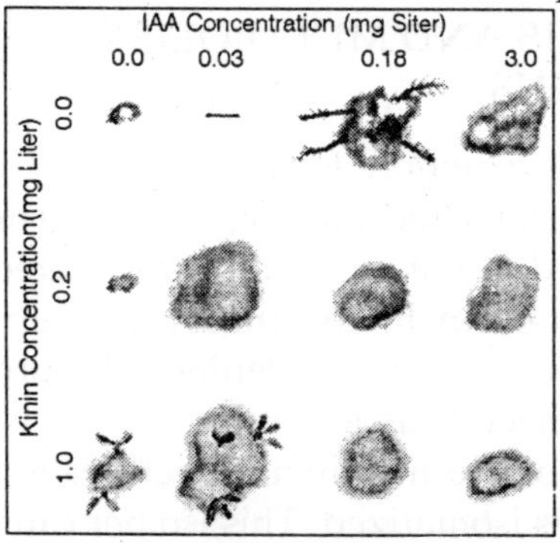

The illustration above shows the effect of cytokinin and auxin concentration on tissue culture experiments.

ETHYLENE

NATURE OF ETHYLENE

Ethylene, unlike the rest of the plant hormone compounds is a gaseous hormone. Like abscisic acid, it is the only member of its class. Of all the known plant growth substance, ethylene has the simplest structure. It is produced in all higher plants and is usually associated with fruit ripening and the tripple response.

$$H_2C{=}CH_2$$

HISTORY OF PLANTS

Ethylene has been used in practice since the ancient Egyptians, who would gas figs in order to stimulate ripening. The ancient Chinese would burn incense in closed rooms to enhance the ripening of pears. It was in 1864, that leaks of gas from street lights showed stunting of growth, twisting of plants, and abnormal thickening of stems (the triple response).

In 1901, a russian scientist named Dimitry Neljubow showed that the active component was ethylene. Doubt discovered that ethylene stimulated abscission in 1917. It wasn't until 1934 that Gane reported that plants synthesize ethylene.

In 1935, Crocker proposed that ethylene was the plant hormone responsible for fruit ripening as well as inhibition of vegetative tissues. Ethylene is now known to have many other functions as well.

Biosynthesis and Metabolism

Ethylene is produced in all higher plants and is produced from methionine in essentially all tissues. Production of ethylene varies with the type of tissue, the plant species, and also the stage of development. The mechanism by which ethylene is produced from methionine is a 3 step process. ATP is an essential component in the synthesis of ethylene from methionine. ATP and water are added to methionine resulting in loss of the three phosphates and S-adenosyl methionine. 1-amino-cyclopropane-1-carboxylic acid synthase facilitates the production of ACC from SAM.

Oxygen is then needed in order ro oxidize ACC and produce ethylene. This reaction is catalyzed by an oxidative enzyme called ethylene forming enzyme. The control of ethylene production has received considerable study.

Study of ethylene has focused around the synthesis promoting effects of auxin, wounding, and drought as well as aspects of fruit-ripening. ACC synthase is the rate limiting step for ethylene production and it is this enzyme that is manipulated in biotechnology to delay fruit ripening in the"flavour saver" tomatoes.

Functions of Ethylene

Ethylene is known to affect the following plant processes:

- Stimulates the release of dormancy.
- Stimulates shoot and root growth and differentiation (triple response)
- May have a role in adventitious root formation.
- Stimulates leaf and fruit abscission.
- Stimulates Bromiliad flower induction.
- Induction of femaleness in dioecious flowers.
- Stimulates flower opening.
- Stimulates flower and leaf senescence.

GIBBERELLINS

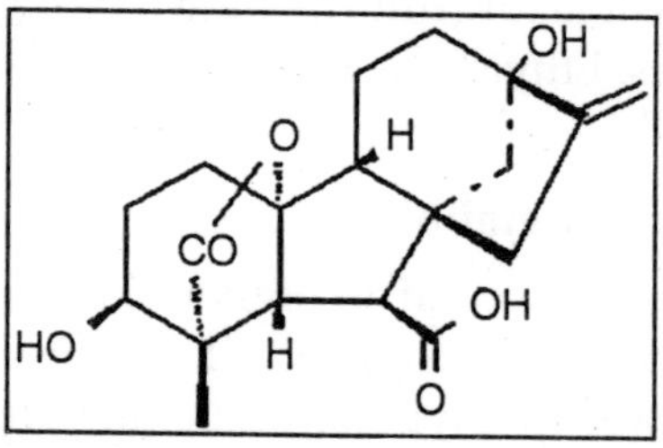

NATURE OF GIBBERELLINS

Unlike the classification of auxins which are classified on the basis of function, gibberellins are classified on the basis of structure as well as function. All gibberellins are derived from the ent-gibberellane skeleton. The structure of this skeleton derivative along with the structure of a few of the active gibberellins are shown above. The gibberellins are named GA1....GAn in order of discovery. Gibberellic acid, which was the first gibberellin to be structurally characterised, is GA3. There are currently 136 GAs identified from plants, fungi and bacteria GA's are widespread and so far ubiquitous in both flowering (angiosperms) and non-flowering (gymnosperms) plants as well as ferns.

GIBBERELLIN BIOSYNTHESIS AND METABOLISM

Gibberellins are diterpenes synthesized from acetyl CoA

via the mevalonic acid pathway. They all have either 19 or 20 carbon units grouped into either four or five ring systems. The fifth ring is a lactone ring as shown in the structures above attached to ring A. Gibberellins are believed to be synthesized in young tissues of the shoot and also the developing seed. It is uncertain whether young root tissues also produce gibberellins.

There is also some evidence that leaves may be the source of some biosynthesis. The pathway by which gibberellins are formed is outlined below and illustrated in figure. 3 acetyl CoA molecules are oxidized by 2 NADPH molecules to produce 3 CoA molecules as a side product and mevalonic acid. Mevalonic acid is then Phosphorylated by ATP and decarboxylated to form isopentyl pyrophosphate. 4 of these molecules form geranylgeranyl pyrophosphate which serves as the donor for all GA carbon atoms.

This compound is then converted to copalylpyrophosphate which has 2 ring systems Copalylpyrophosphate is then converted to kaurene which has 4 ring systems. Subsequent oxidations reveal kaurenol (alcohol form), kaurenal (aldehyde form), and kaurenoic acid respectively.

Kaurenoic acid is converted to the aldehyde form of GA12 by decarboxylation. GA12 is the 1st true gibberellane ring system with 20 carbons. From the aldehyde form of GA12 arise both 20 and 19 carbon gibberellins but there are many mechanisms by which these other compounds arise.

Certain commercial chemicals which are used to stunt growth do so in part because they block the synthesis of gibberellins. Some of these chemicals are Phosphon D, Amo-1618, Cycocel (CCC), ancymidol, and paclobutrazol. During active growth, the plant will metabolize most gibberellins by hydroxylation to inactive conjugates quickly with the exception of GA3. GA3 is degraded much slower which helps to explain why the symptoms initially associated with the hormone in the disease bakanae are present. Inactive conjugates might be stored or translocated via the phloem and xylem before their release (activation) at the proper time and in the proper tissue.

FUNCTIONS OF GIBBERELLINS

Active gibberellins show many physiological effects, each depending on the type of gibberellin present as well as the species of plant.

Some of the physiological processes stimulated by gibberellins are outlined below:

- Stimulate stem elongation by stimulating cell division and elongation.
- Stimulates bolting/flowering in response to long days.
- Breaks seed dormancy in some plants which require stratification or light to induce germination.
- Stimulates enzyme production (a-amylase) in germinating cereal grains for mobilization of seed reserves.
- Induces maleness in dioecious flowers (sex expression).
- Can cause parthenocarpic (seedless) fruit development.
- Can delay senescence in leaves and citrus fruits.

GROWTH AND DEVELOPMENT

Growth is an increase in size of basic parts of plants (due to production of extra cells), whereas development is the production of new structures in the progress towards maturity - flowers, fruits etc.

PLANT HORMONES VS ANIMAL HORMONES

In that green plants make growth in response to stimuli from their surroundings, they show sensitivity, *i.e.* one of the seven characteristic processes of living organisms. However, green plants respond to fewer stimuli than animals, and they make a slower and more restricted response.

The 3 stimuli to which green plants responds are:

1. Light.
2. Gravity.
3. Water/touch (peasticks etc)NOT CO_2.

If a plant responds by growing in a particular direction, it is called a tropism. Positive tropisms are towards the stimulus, negative tropisms are away from the stimulus.

REGULATION OF DEVELOPMENT

Stimulation of Roots in Cuttings

When cuttings are taken, the shoot which is selected is removed from the plant and disconnected from a supply of water and minerals from the roots. It is therefore important to encourage the development of roots, to aid the establishment of the"cutting" as an independent plant. "Hormone rooting powders" contain the substance indoleacetic acid IAA which has been identified as the same chemical as"auxin" which causes curvature in roots and shoots. Usually the newly exposed surface of the cutting is dipped into a powder containing the hormone, mixed with inert powder, and then it is placed into a pot containing moist compost or soil, which is gently firmed around the cutting and placed in a cool moist environment. There are a variety of other chemically similar compounds which may be used for this purpose.

HORMONE WEEDKILLERS

Although there are a wide variety of chemicals which can be toxic to plants or which damage their exposed parts, such as leaves and shoots, these are of limited use due to problems of toxicity to humans and other animals, and danger due to their oxidising nature. These substances may not affect parts of plants which are underground, and there is a likelihood that new growth will sprout again from these parts. There are also many situations in which more specificity is required: killing some plants but not others, and leaving soil fit for subsequent growth of plants which are required.

Hormone weedkillers have a variety of advantages over more traditional substances:

- They should be more toxic to plants than animals.
- They should be carried to underground plants and kill then too.
- Some weedkillers have a greater effect on"broad-leaved weeds" than on grasses and also cereal crops (monocots). For this reason they are called selective weedkillers.

These may be used to control weeds amongst the grass on lawns, or to reduce competition from weeds in cereal crops. Examples include compounds with complicated chemical names, popularly abbreviated to 2.4-D and 2,4,5-T. Other compounds such as Dalapon are more effective against monocots than dicots, so they could be used in the opposite situation. In fact these weedkilllers cause sensitive plants to grow too fast, damaging their cells so that their conducting tissue does not develop properly, or they use up their food reserves and die.

Weed

A plant growing in the wrong place - *i.e.* not planted by a gardener or grower. In other words, a plant which is able to compete and grow well in the prevailing conditions - probably in its own natural habitat!

Consequences of Using Selective Weedkillers

It must be admitted that the widespread use of selective weedkillers in European and Western agriculture has improved the efficiency of crop production (productivity of arable farms), in response to policies promoted by government ministries. This has been quite successful. resulting in the excessive production of cereal and other crops - so called"grain mountains".

However this apparent economic improvement has been at the expense of biodiversity in the countryside, as many"weed" species are food sources and act as shelter for many species of wildlife. Many plants which used to flower alongside crops and weeds, adding to the visual impact of the countryside, have been adversely affected by herbicides sprayed onto fields. Similarly, the survival of animal species, such as birds and butterflies, which add immeasurably to the public enjoyment of the wider environment, are consequently threatened more by the lack of food and protection than by the killing power of the chemicals involved.

Recently, there have been a series of developments aimed at reversing this trend, including"set-aside" where farmers are paid compensation in order to leave sections of their farms fallow in order to revert to"natural" conditions which should support normal fauna and flora. Regulations about maintenance of hedges and borders around fields have also been relaxed in order to increase the areas which can act as habitats for wildlife.

Development of Fruits

The ripening of fruits is known to be affected by a number of plant hormones. Commercially, growers sometimes use chemical sprays to prevent fruit from falling from trees. The substances used are auxins such as IAA or substances such as 2,4-D which is also used as a herbicide. Depending on weather conditions, this normally happens with a proportion of the fruit crop, leading to uncertainty as to the likely productivity of the crop. The consistency of cropping and ease of picking can thus be increased. Normally, a flower which is pollinated develops into a fruit, containing seeds derived from the fertilised egg cell. However, in some circumstances, fruit develops without pollination taking place. In seedless varieties of fruits this parthenocarpic fruit development occurs when other parts of the fruit produce auxins which would otherwise come from the seeds.

Other chemical substances acting as growth regulators can be used to induce flowering and alter the number and shape of fruit forming. The gas ethane (ethylene) has also been shown to have a hormone-like activity on fruit and is used to control the ripening of stored fruit such as bananas and apples which are best picked and shipped before maturity.

8

DNA and RNA

The nucleic acids are the building blocks of living organisms. You may have heard of DNA described the same way. Guess what? DNA is just one type of nucleic acid. Some other types are RNA, mRNA, and tRNA. All of these "Na's" work together to help cells replicate and build proteins.

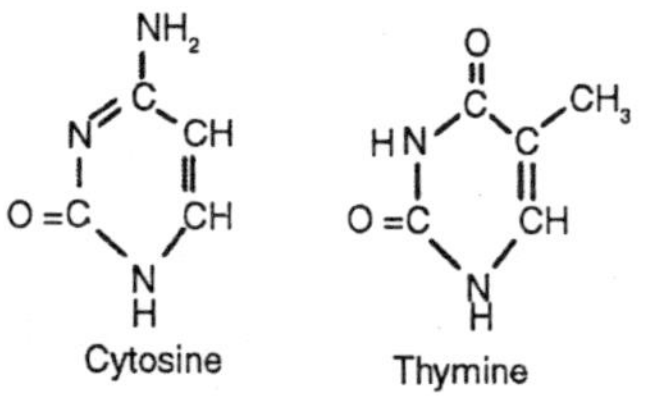

Pyrimidine Bases

BASICS

We already told you about the biggie nucleic acids (DNA, RNA, mRNA, tRNA). They are actually made up chains of base pairs stretching from only a few to millions. When those pairs combine in super long chains (DNA), they make a shape called a double helix. The double helix shape is like a twisty ladder. The base pairs are the rungs. We're very close to talking about the biology of cells here. Back to the chemistry..

FIVE EASY PIECES

There are five easy parts of nucleic acids. All nucleic acids are made up of the same building blocks (monomers). Chemists call the monomers nucleotides. The five pieces are Uracil,Cytosine, Thymine, Adenine, and Guanine. Just as there are twenty (20) amino acids needed by humans to survive, there are five (5)nucleotides.

These nucleotides are made of three parts:

1. A five carbon sugar
2. A base that has a nitrogen (N) atom
3. An ion of phosphoric acid

The first isolation of what we now refer to as DNA was accomplished by Johann Friedrich Miescher *circa* 1870. He reported finding a weakly acidic substance of unknown function in the nuclei of human white blood cells, and named this material "nuclein". A few years later, Miescher separated nuclein into protein and nucleic acid components. In the 1920's nucleic acids were found to be major components of chromosomes, small gene-carrying bodies in the nuclei of complex cells. Elemental analysis of nucleic acids showed the presence of phosphorus, in addition to the usual C, H, N & O. Unlike proteins, nucleic acids contained no sulfur. Complete hydrolysis of chromosomal nucleic acids gave inorganic phosphate, 2-deoxyribose (a previously unknown sugar) and four different heterocyclic bases (shown in the following diagram). To reflect the unusual sugar component, chromosomal nucleic acids are called deoxyribonucleic acids, abbreviated DNA. Analogous nucleic acids in which the sugar component is ribose are termed ribonucleic acids, abbreviated RNA. The acidic character of the nucleic acids was attributed to the phosphoric acid moiety.

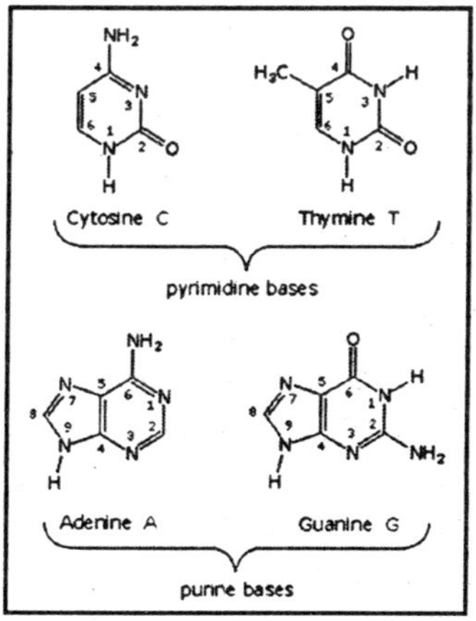

The two monocyclic bases shown here are classified as pyrimidines, and the two bicyclic bases are purines. Each has at least one N-H site at which an organic substituent may be attached. They are all polyfunctional bases, and may exist in tautomeric forms.

Base-catalyzed hydrolysis of DNA gave four nucleoside products, which proved to be N-glycosides of 2'-deoxyribose combined with the heterocyclic amines. Structures and names for these nucleosides will be displayed above by clicking on the heterocyclic base diagram. The base components are coloured green, and the sugar is black. As noted in the 2'-deoxycytidine structure on the left, the numbering of the sugar carbons makes use of primed numbers to distinguish them from the heterocyclic base sites. The corresponding N-glycosides of the common sugar ribose are the building blocks of RNA, and are named adenosine, cytidine, guanosine and uridine (a thymidine analog missing the methyl group).

From this evidence, nucleic acids may be formulated as alternating copolymers of phosphoric acid (P) and nucleosides (N), as shown:

$$\sim P - N - P - N' - P - N'' - P - N''' - P - N \sim$$

At first the four nucleosides, distinguished by prime marks in this crude formula, were assumed to be present in equal amounts, resulting in a uniform structure, such as that of starch. However, a compound of this kind, presumably common to all organisms, was considered too simple to hold the hereditary information known to reside in the chromosomes. This view was challenged in 1944, when Oswald Avery and colleagues demonstrated that bacterial DNA was likely the genetic agent that carried information from one organism to another in a process called "transformation".

He concluded that *"nucleic acids must be regarded as possessing biological specificity, the chemical basis of which is as yet undetermined."* Despite this finding, many scientists continued to believe that chromosomal proteins, which differ across species, between individuals, and even within a given organism, were the locus of an organism's genetic information.

It should be noted that single celled organisms like bacteria do not have a well-defined nucleus. Instead, their single chromosome is associated with specific proteins in a region called a "nucleoid". Nevertheless, the DNA from bacteria has the same composition and general structure as that from multicellular organisms, including human beings.

Views about the role of DNA in inheritance changed in the late 1940's and early 1950's. By conducting a careful analysis of DNA from many sources, Erwin Chargaff found its composition to be species specific. In addition, he found that the amount of adenine (A) always equaled the amount of thymine (T), and the amount of guanine (G) always equaled the amount of cytosine (C), regardless of the DNA source. As set forth in the following table, the ratio of (A+T) to (C+G) varied from 2.70 to 0.35. The last two organisms are bacteria.

Nucleoside Base Distribution in DNA

Organism	Base Composition (mole %)				Base Ratios		Ratio (A+T)/(G+C)
	A	G	T	C	A/T	G/C	
Human	30.9	19.9	29.4	19.8	1.05	1.00	1.52
Chicken	28.8	20.5	29.2	21.5	1.02	0.95	1.38
Yeast	31.3	18.7	32.9	17.1	0.95	1.09	1.79
Clostridium perfringens	36.9	14.0	36.3	12.8	1.01	1.09	2.70
Sarcina lutea	13.4	37.1	12.4	37.1	1.08	1.00	0.35

In a second critical study, Alfred Hershey and Martha Chase showed that when a bacterium is infected and genetically transformed by a virus, at least 80% of the viral DNA enters the bacterial cell and at least 80% of the viral protein remains outside. Together with the Chargaff findings this work established DNA as the repository of the unique genetic characteristics of an organism.

THE CHEMICAL NATURE OF DNA

The polymeric structure of DNA may be described in terms of monomeric units of increasing complexity. In the top shaded box of the following illustration, the three relatively simple components mentioned earlier are shown. Below that on the left, formulas for phosphoric acid and a nucleoside are drawn.Condensation polymerization of these leads to the DNA formulation outlined above. Finally, a 5'- monophosphate ester, called a nucleo-tide may be drawn as a single monomer unit, shown in the shaded box to the right. Since a monophosphate ester of this kind is a strong acid (pK_a of 1.0), it will be fully ionized at the usual physiological pH (ca.7.4).

Names for these DNA components are given in the table to the right of the diagram. Isomeric 3'-monophospate nucleotides are also known, and both isomers are found in cells. They may be obtained by selective hydrolysis of DNA through the action of nuclease enzymes. Anhydride-like di- and tri-phosphate nucleotides have been identified as important energy carriers in biochemical reactions, the most common being ATP (adenosine 5'-triphosphate).

Names of DNA Base Derivatives

Base	Nucleoside	5'-Nucleotide
Adenine	2'-Deoxyadenosine	2'-Deoxyadenosine-5'-monophosphate
Cytosine	2'-Deoxycytidine	2'-Deoxycytidine-5'-monophosphate
Guanine	2'-Deoxyguanosine	2'-Deoxyguanosine-5'-monophosphate
Thymine	2'-Deoxythymidine	2'-Deoxythymidine-5'-monophosphate

A complete structural representation of a segment of the DNA polymer formed from 5'-nucleotides. Several important characteristics of this formula should be noted.

- First, the remaining P-OH function is quite acidic and is completely ionized in biological systems.
- Second, the polymer chain is structurally directed. One end is different from the other.
- Third, although this appears to be a relatively simple polymer, the possible permutations of the four nucleosides in the chain become very large as the chain lengthens.
- Fourth, the DNA polymer is much larger than originally believed. Molecular weights for the DNA from multicellular organisms are commonly 10^9 or greater.

Information is stored or encoded in the DNA polymer by the pattern in which the four nucleotides are arranged. To access this information the pattern must be "read" in a linear fashion, just as a bar code is read at a supermarket checkout. Because living organisms are extremely complex, a correspondingly large amount of information related to this complexity must be stored in the DNA.

Consequently, the DNA itself must be very large, as noted above. Even the single DNA molecule from an *E. coli* bacterium is found to have roughly a million nucleotide units in a polymer strand, and would reach a millimeter in length if stretched out. The nuclei of multicellular organisms incorporate chromosomes, which are composed of DNA combined with nuclear proteins called histones. The fruit fly has 8 chromosomes, humans have 46 and dogs 78 (note that the amount of DNA in a cell's nucleus does not correlate with the number of chromosomes). The DNA from the smallest human chromosome is over ten times larger than *E. coli* DNA, and it has been estimated that the total DNA in a human cell would extend to 2 meters in length if unraveled. Since the nucleus is only about 5ìm in diameter, the chromosomal DNA must be packed tightly to fit in that small volume.

In addition to its role as a stable informational library, chromosomal DNA must be structured or organized in such a way that the chemical machinery of the cell will have easy access to that information, in order to make important molecules such as polypeptides. Furthermore, accurate copies of the DNA code must be created as cells divide, with the replicated DNA molecules passed on to subsequent cell generations, as well as to progeny of the organism.

RNA, A DIFFERENT NUCLEIC ACID

The high molecular weight nucleic acid, DNA, is found chiefly in the nuclei of complex cells, known as eucaryotic cells, or in the nucleoid regions ofprocaryotic cells, such as bacteria. It is often associated with proteins that help to pack it in a usable fashion. In contrast, a lower molecular weight, but much more abundant nucleic acid, RNA, is distributed throughout the cell, most commonly in small numerous organelles called ribosomes. Three kinds of RNA are identified, the largest subgroup (85 to 90%) being ribosomal RNA, rRNA, the major component of ribosomes, together with proteins. The size of rRNA molecules varies, but is generally less than a thousandth the size of DNA. The other forms of RNA are messenger RNA, mRNA, and transfer RNA, tRNA. Both have a more transient existence and are smaller than rRNA.

All these RNA's have similar constitutions, and differ from DNA in two important respects. As shown in the following diagram, the sugar component of RNA is ribose, and the pyrimidine base uracil replaces the thymine base of DNA. The RNA's play a vital role in the transfer of information (transcription) from the DNA library to the protein factories called ribosomes, and in the interpretation of that information (translation) for the synthesis of specific polypeptides. These functions will be described later. A complete structural representation of a segment of the RNA polymer formed from 5'-nucleotides.

THE SECONDARY STRUCTURE OF DNA

In the early 1950's the primary structure of DNA was well established, but a firm understanding of its secondary structure was lacking. Indeed, the situation was similar to that occupied by the proteins a decade earlier, before the alpha helix and pleated sheet structures were proposed by Linus Pauling. Many researchers grappled with this problem, and it was generally conceded that the molar equivalences of base pairs (A & T and C & G) discovered by Chargaff would be an important factor. A scientist, working at King's College, obtained X-ray diffraction evidence that suggested a long helical structure of uniform thickness.

THE DOUBLE HELIX

After many trials and modifications, an ingenious double helix model for the secondary structure of DNA. Two strands of DNA were aligned anti-parallel to each other, *i.e.* with opposite 3' and 5' ends, as shown in part a of the following diagram. Complementary primary nucleotide structures for each strand allowed intra-strand hydrogen bonding between each pair of bases. These complementary strands are coloured red and green in the diagram. Coiling these coupled strands then leads to a double helix structure, shown as cross-linked ribbons in part b of the diagram. A space-filling molecular model of a short segment is displayed in part c on the right.

The helix shown here has ten base pairs per turn, and rises 3.4 Å in each turn. This right-handed helix is the favoured conformation in aqueous systems, and has been termed the B-helix. Two alternating grooves are evident, a wide and deep major groove, and a shallow and narrow minor groove. Other molecules, including polypeptides, may insert into these grooves, and in so doing perturb the chemistry of DNA. Other helical structures of DNA have also been observed, and are designated by letters (*e.g.* A and Z).

The Double Helix Structure for DNA

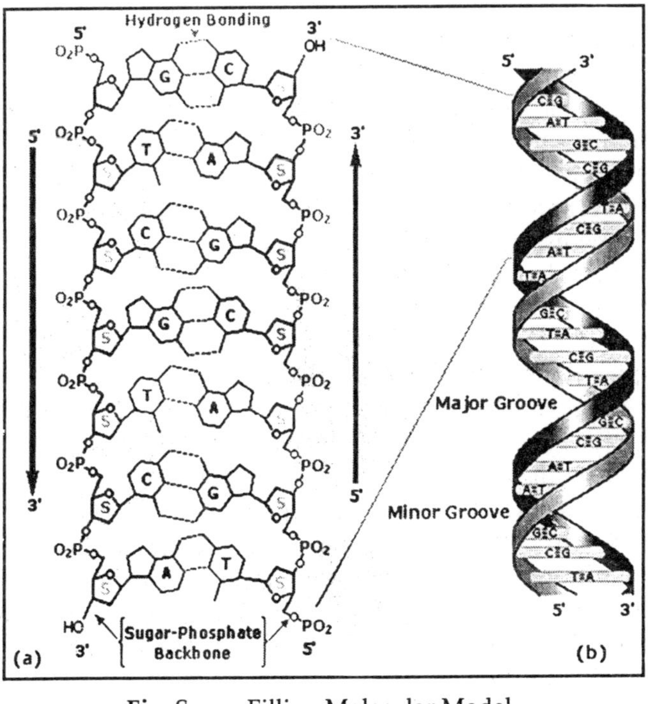

Fig. Space-Filling Molecular Model.

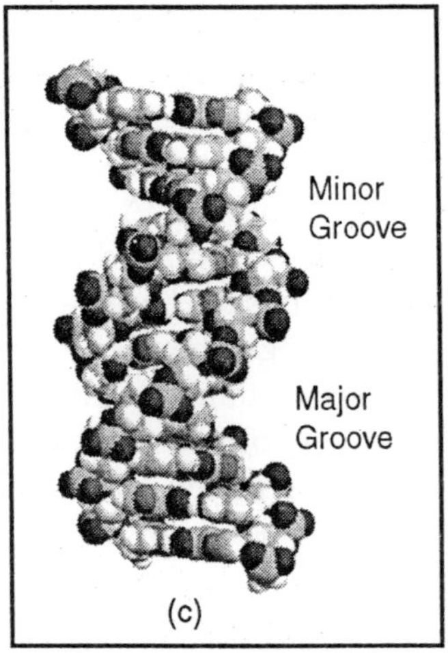

Fig. Model of a Short DNA Segment

DNA Replication

In their announcement of a double helix structure for DNA, It is stated, *"It has not escaped our notice that the specific pairing we have postulated immediately suggests a possible copying mechanism for the genetic material."*. The essence of this suggestion is that, if separated, each strand of the molecule might act as a template on which a new complementary strand might be assembled, leading finally to two identical DNA molecules. Indeed, replication does take place in this fashion when cells divide, but the events leading up to the actual synthesis of complementary DNA strands are sufficiently complex that they will not be described in any detail.

As depicted in the following drawing, the DNA of a cell is tightly packed into chromosomes. First, the DNA is wrapped around small proteins called histones (coloured pink below). These bead-like structures are then further organized and folded into chromatin aggregates that make up the chromosomes. An overall packing efficiency of 7,000 or more is thus achieved. Clearly a sequence of unfolding events must take place before the information encoded in the DNA can be used or replicated.

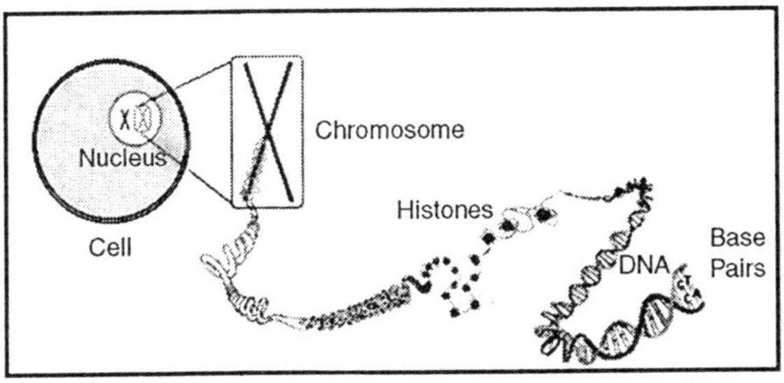

Once the double stranded DNA is exposed, a group of enzymes act to accomplish its replication.

These are described briefly here:

- *Topoisomerase:* This enzyme initiates unwinding of the double helix by cutting one of the strands.
- *Helicase:* This enzyme assists the unwinding. Note that many hydrogen bonds must be broken if the strands are to be separated..
- *SSB:* A single-strand binding-protein stabilizes the separated strands, and prevents them from recombining, so that the polymerization chemistry can function on the individual strands.
- *DNA Polymerase:* This family of enzymes link together nucleotide triphosphate monomers as they hydrogen bond to complementary bases. These enzymes also check for errors (roughly ten per billion), and make corrections.
- *Ligase:* Small unattached DNA segments on a strand are united by this enzyme.

Polymerization of nucleotides takes place by the phosphorylation reaction described by the following equation.

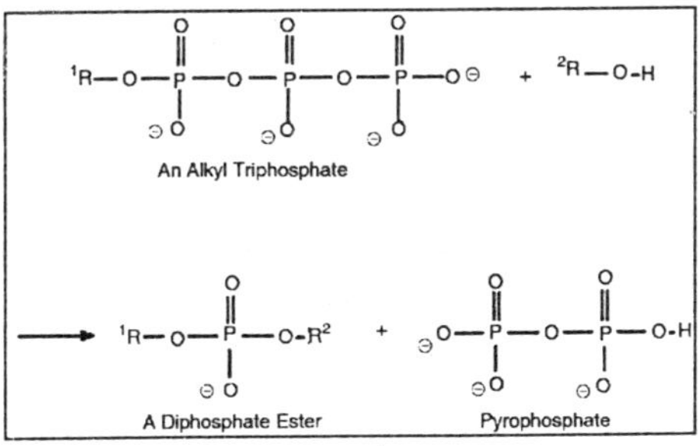

Di- and triphosphate esters have anhydride-like structures and are consequently reactive phosphorylating reagents, just as carboxylic anhydrides are acylating reagents. Since the pyrophosphate anion is a better leaving group than phosphate, triphosphates are more powerful phosphorylating agents than are diphosphates.

Formulas for the corresponding 5'-derivatives of adenosine will be displayed, and similar derivatives exist for the other three common nucleosides. The DNA polymerization process that builds the complementary strands in replication, could in principle take place in two ways. Referring to the general equation above, R^1 could represent the next nucleotide unit to be attached to the growing DNA strand, with R^2 being this strand. Alternatively, these assignments could be reversed.

In practice, the former proves to be the best arrangement. Since triphosphates are very reactive, the lifetime of such derivatives in an aqueous environment is relatively short. However, such derivatives of the individual nucleosides are repeatedly synthesized by the cell for a variety of purposes, providing a steady supply of these reagents.

In contrast, the growing DNA segment must maintain its functionality over the entire replication process, and can not afford to be changed by a spontaneous hydrolysis event. As a result, these chemical properties are best accomodated by a polymerization process that proceeds at the 3'-end of the growing strand by 5'-phosphorylation involving a nucleotide triphosphate.

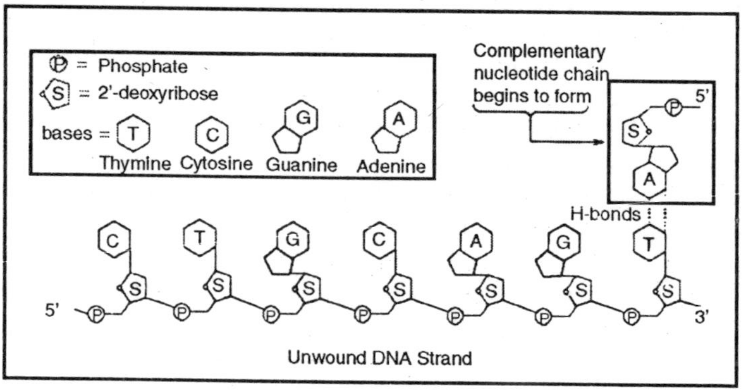

The polymerization mechanism described here is constant. It always extends the developing DNA segment towards the 3'-end (*i.e.* when a nucleotide triphosphate attaches to the free 3'-hydroxyl group of the strand, a new 3'-hydroxyl is generated). There is sometimes confusion on this point, because the original DNA strand that serves as a template is read from the 3'-end towards the 5'-end, and authors may not be completely clear as to which terminology is used.

Because of the directional demand of the polymerization, one of the DNA strands is easily replicated in a continuous fashion, whereas the other strand can only be replicated in short segmental pieces. This is illustrated in the following diagram. Separation of a portion of the double helix takes place at a site called the replication fork. As replication of the separate strands occurs, the replication fork moves away (to the left in the diagram), unwinding additional lengths of DNA. Since the fork in the diagram is moving towards the 5'-end of the red-coloured strand, replication of this strand may take place in a continuous fashion (building the new green strand in a 5' to 3' direction). This continuously formed new strand is called the leading strand. In contrast, the replication fork moves towards the 3'-end of the original green strand, preventing continuous polymerization of a complementary new red strand. Short segments of complementary DNA, called Okazaki fragments, are produced, and these are linked together later by the enzyme ligase. This new DNA strand is called the lagging strand.

When you consider that a human cell has roughly 10^9 base pairs in its DNA, and may divide into identical daughter cells in 14 to 24 hours, the efficiency of DNA replication must be extraordinary. The procedure described above will replicate about 50 nucleotides per second, so there must be many thousand such replication sites in action during cell division. A given length of double stranded DNA may undergo strand unwinding at numerous sites in response to promoter actions. The unraveled "bubble" of single stranded DNA has two replication forks, so assembly of new complementary strands may proceed in two directions. The polymerizations associated with several such bubbles fuse together to achieve full replication of the entire DNA double helix. Note that the events shown proceed from top to bottom in the diagram.

Repair of DNA Damage and Replication Errors

One of the benefits of the double stranded DNA structure is that it lends itself to repair, when structural damage or replication errors occur.

Several kinds of chemical change may cause damage to DNA:

- Spontaneous hydrolysis of a nucleoside removes the heterocyclic base component.
- Spontaneous hydrolysis of cytosine changes it to a uracil.
- Various toxic metabolites may oxidize or methylate heterocyclic base components.
- Ultraviolet light may dimerize adjacent cytosine or thymine bases.

All these transformations disrupt base pairing at the site of the change, and this produces a structural deformation in the double helix.. Inspection-repair enzymes detect such deformations, and use the undamaged nucleotide at that site as a template for replacing the damaged unit. These repairs reduce errors in DNA structure from about one in ten million to one per trillion.

RNA AND PROTEIN SYNTHESIS

The genetic information stored in DNA molecules is used as a blueprint for making proteins. Why proteins? Because these macromolecules have diverse primary, secondary and tertiary structures that equip them to carry out the numerous functions necessary to maintain a living organism.

As noted in the protein, these functions include:

- Structural integrity (hair, horn, eye lenses etc.).
- Molecular recognition and signaling (antibodies and hormones).
- Catalysis of reactions (enzymes)..
- Molecular transport (hemoglobin transports oxygen).
- Movement (pumps and motors).

The critical importance of proteins in life processes is demonstrated by numerous genetic diseases, in which small modifications in primary structure produce debilitating and often disasterous consequences.

Such genetic diseases incluse Tay-Sachs, phenylketonuria (PKU), sickel cell anemia, achondroplasia, and Parkinson disease.

The unavoidable conclusion is that proteins are of central importance in living cells, and that proteins must therefore be continuously prepared with high structural fidelity by appropriate cellular chemistry.

Early geneticists identifed genes as hereditary units that determined the appearance and/ or function of an organism (*i.e.* its phenotype). We now define genes as sequences of DNA that occupy specific locations on a chromosome.

The original proposal that each gene controlled the formation of a single enzyme has since been modified as: one gene = one polypeptide. The intriguing question of how the information encoded in DNA is converted to the actual construction of a specific polypeptide has been the subject of numerous studies, which have created the modern field of Molecular Biology.

CENTRAL DOGMA AND TRANSCRIPTION

It is proposed that information flows from DNA to RNA in a process called transcription, and is then used to synthesize polypeptides by a process called translation. Transcription takes place in a manner similar to DNA replication.

A characteristic sequence of nucleotides marks the beginning of a gene on the DNA strand, and this region binds to a promoter protein that initiates RNA synthesis. The double stranded structure unwinds at the promoter site., and one of the strands serves as a template for RNA formation, as depicted in the following diagram.

The RNA molecule thus formed is single stranded, and serves to carry information from DNA to the protein synthesis machinery called ribosomes. These RNA molecules are therefore called messenger-RNA (mRNA).

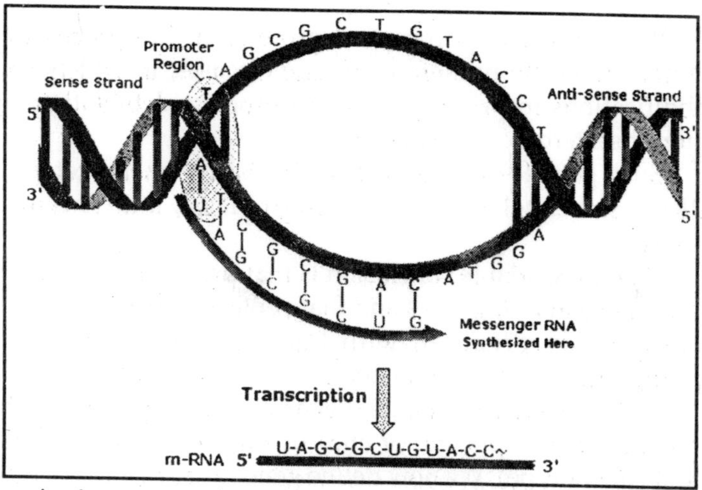

An important distinction must be made here. One of the DNA strands in the double helix holds the genetic information used for protein synthesis. This is called the sense strand, or information strand. The complementary strand that binds to the sense strand is called the anti-sense strand and it serves as a template for generating a mRNA molecule that delivers a copy of the sense strand information to a ribosome.

The promoter protein binds to a specific nucleotide sequence that identifies the sense strand, relative to the anti-sense strand. RNA synthesis is then initiated in the 3' direction, as nucleotide triphosphates bind to complementary bases on the template strand, and are joined by phosphate diester linkages. An animation of this process for DNA replication was presented earlier. A characteristic "stop sequence" of nucleotides terminates the RNA synthesis. The messenger molecule (coloured orange above) is released into the cytoplasm to find a ribosome, and the DNA then rewinds to its double helix structure.

In eucaryotic cells the initially transcribed m-RNA molecule is usually modified and shortened by an "editing" process that removes irrelevent material. The DNA of such organisms is often thousands of times larger and more complex than that composing the single chromosome of a procaryotic bacterial cell. This difference is due in part to repetitive nucleotide sequences (ca. 25% iu the human genome). Furthermore, over 95% of human DNA is found in intervening sequences that separate genes and parts of genes. The informational DNA segments that make up genes are called exons, and the noncoding segments are called introns.

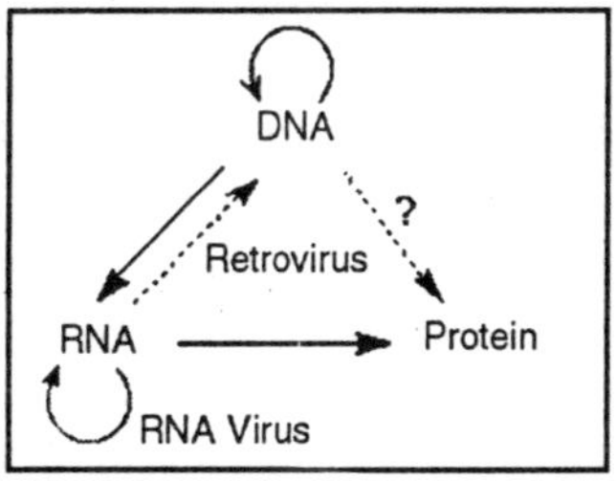

Before the mRNA molecule leaves the nucleus, the nonsense bases that make up the introns are cut out, and the informationally useful exons are joined together in a step known as RNA splicing. In this fashion shorter mRNA molecules carrying the blueprint for a specific protein are sent on their way to the ribosome factories.

The Central Dogma of molecular biology, which at first was formulated as a simple linear progression of information from DNA to RNA to Protein, is summarized in the following illustration. The replication process on the left consists of passing information from a parent DNA molecule to daughter molecules. The middle transcription process copies this information to a mRNA molecule. Finally, this information is used by the chemical machinery of the ribosome to make polypeptides.

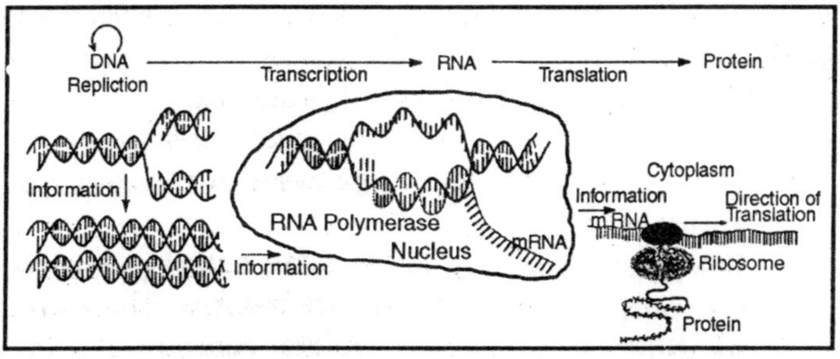

As more has been learned about these relationships, the central dogma has been refined to the representation displayed on the right. The dark blue arrows show the general, well demonstrated, information transfers noted above. It is now known that an RNA-dependent DNA polymerase enzyme, known as a reverse transcriptase, is able to transcribe a single-stranded RNA sequence into double-stranded DNA (magenta arrow).

Such enzymes are found in all cells and are an essential component of retroviruses (*e.g.* HIV), which require RNA replication of their genomes (green arrow). Direct translation of DNA information into protein synthesis (orange arrow) has not yet been observed in a living organism. Finally, proteins appear to be an informational dead end, and do not provide a structural blueprint for either RNA or DNA. In the following section the last fundamental relationship, that of structural information translation from mRNA to protein, will be described

TRANSLATION

Translation is a more complex process than transcription. This would, of course, be expected. After all, the coded messages produced by the Enigma machine could be copied easily, but required a considerable decoding effort before they could be read with understanding.

In a similar sense, DNA replication is simply a complementary base pairing exercise, but the translation of the four letter (bases) alphabet code of RNA to the twenty letter (amino acids) alphabet of protein literature is far from trivial. Clearly, there could not be a direct one-to-one correlation of bases to amino acids, so the nucleotide letters must form short words or codons that define specific amino acids.

Many questions pertaining to this genetic code were posed:

- How many RNA nucleotide bases designate a specific amino acid?
  If separate groups of nucleotides, called codons, serve this purpose, at least three are needed. There are $4^3 =$ 64 different nucleotide triplets, compared with $4^2 =$ 16 possible pairs.
- Are the codons linked separately or do they overlap?

Sequentially joined triplet codons will result in a nucleotide chain three times longer than the protein it describes. If overlapping codons are used then fewer total nucleotides would be required.

- If triplet segments of mRNA designate specific amino acids in the protein, how are the codons identified? For the sequence ~CUAGGU~ are the codons CUA & GGU or ~C, UAG & GU~ or ~CU, AGG & U~?
- Are all the codon words the same size?

In Morse code the most widely used letters are shorter than less common letters. Perhaps nature employs a similar scheme.

Physicists and mathematicians, as well as chemists and microbiologists all contributed to unravelling the genetic code. Although earlier proposals assumed efficient relationships that correlated the nucleotide codons uniquely with the twenty fundamental amino acids, it is now apparent that there is considerable redundancy in the code as it now operates. Furthermore, the code consists exclusively of non-overlapping triplet codons.

Clever experiments provided some of the earliest breaks in deciphering the genetic code. It is found that RNA from many different organisms could initiate specific protein synthesis when combined with broken E.coli cells (the enzymes remain active).

A synthetic polyuridine RNA induced synthesis of poly-phenylalanine, so the UUU codon designated phenylalanine. Likewise an alternating ~CACA~ RNA led to synthesis of a ~His-Thr-His-Thr~ polypeptide.

The following table presents the present day interpretation of the genetic code. Note that this is the RNA alphabet, and an equivalent DNA codon table would have all the U nucleotides replaced by T.

Methionine and tryptophan are uniquely represented by a single codon. At the other extreme, leucine is represented by eight codons. The average redundancy for the twenty amino acids is about three. Also, there are three stop codons that terminate polypeptide synthesis.

RNA Codons for Protein Synthesis

First Position		Second Position: U	C	A	G	Third Position
	U	UUU Phe [F]	UCU Ser [S]	UAU Tyr [Y]	UGU Cys [C]	U
		UUC Phe [F]	UCC Ser [S]	UAC Tyr [Y]	UGC Cys [C]	C
		UUA Leu [L]	UCA Ser [S]	UAA Stop	UGA Stop	A
		UUG Leu [L]	UCG Ser [S]	UAG Stop	UGG Trp [W]	G
	C	CUU Leu [L]	CCU Pro [P]	CAU His [H]	CGU Arg [R]	U
		CUC Leu [L]	CCC Pro [P]	CAC His [H]	CGC Arg [R]	C
		CUA Leu [L]	CCA Pro [P]	CAA Gln [Q]	CGA Arg [R]	A
		CUG Leu [L]	CCG Pro [P]	CAG Gln [Q]	CGG Arg [R]	G
	A	AUU Ile [I]	ACU Thr [T]	AAU Asn [N]	AGU Ser [S]	U
		AUC Ile [I]	ACC Thr [T]	AAC Asn [N]	AGC Ser [S]	C
		AUA Ile [I]	ACA Thr [T]	AAA Lys [K]	AGA Arg [R]	A
		AUG Met [M]	ACG Thr [T]	AAG Lys [K]	AGG Arg [R]	G
	G	GUU Val [V]	GCU Ala [A]	GAU Asp [D]	GGU Gly [G]	U
		GUC Val [V]	GCC Ala [A]	GAC Asp [D]	GGC Gly [G]	C
		GUA Val [V]	GCA Ala [A]	GAA Glu [E]	GGA Gly [G]	A
		GUG Val [V]	GCG Ala [A]	GAG Glu [E]	GGG Gly [G]	G

Transfer RNA Molecules

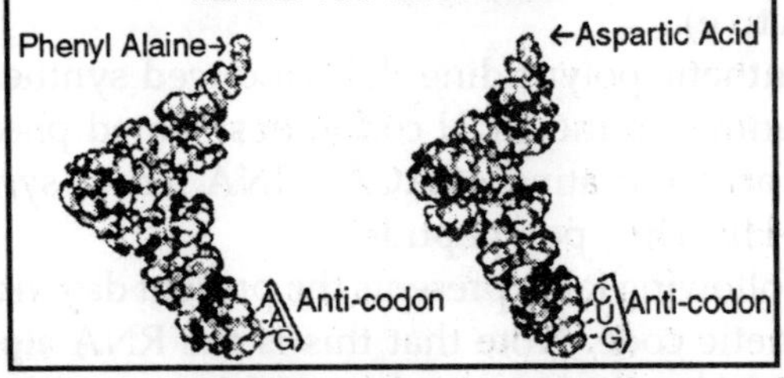

The translation process is fundamentally straightforward. The mRNA strand bearing the transcribed code for synthesis of a protein interacts with relatively small RNA molecules (about 70-nucleotides) to which individual amino acids have been attached by an ester bond at the 3'-end. These transfer RNA's(tRNA) have distinctive three-dimensional structures cosisting of loops of single-stranded RNA connected by double stranded segments. This cloverleaf secondary structure is further wrapped into an "L-shaped" assembly, having the amino acid at the end of one arm, and a characteristic anti-codon region at the other end. The anti-codon consists of a nucleotide triplet that is the complement of the amino acid's codon(s). Models of two such tRNA molecules are shown to the right. When read from the top to the bottom, the anti-codons depicted here should complement a codon in the previous table. A cell's protein synthesis takes place in organelles called ribosomes. Ribosomes are complex structures made up of two distinct and separable subunits (one about twice the size of the other). Each subunit is composed of one or two RNA molecules (60-70%) associated with 20 to 40 small proteins (30-40%). The ribosome accepts a mRNA molecule, binding initially to a characteristic nucleotide sequence at the 5'-end (coloured light blue in the following diagram).

This unique binding assures that polypeptide synthesis starts at the right codon. A tRNA molecule with the appropriate ant-codon then attaches at the starting point and this is followed by a series of adjacent tRNA attachments, peptide bond formation and shifts of the ribosome along the mRNA chain to expose new codons to the ribosomal chemistry.

The following diagram is designed as a slide show illustrating these steps. The outcome is synthesis of a polypeptide chain corresponding to the mRNA blueprint. A "stop codon" at a designated position on the mRNA terminates the synthesis by introduction of a "Release Factor".

DRUG ACTING ON DNA

MECHANISM OF INTERCALATING

Intercalating agents wedge between bases along the DNA. The intercalated drug molecules affect the structure of the DNA, preventing polymerases and other DNA binding proteins from functioning properly. The result is prevention of DNA synthesis, inhibition of transcription and induction of mutations. The effect of principal alkaloids (sanguinarine, chelerythrine, coptisine, chelidonine) of greater celandine Chelidonium majus L., as well as the alkaloids from Colchicum autumnale L.(colchicine and colchamine) on calcium accumulation and oxidative phosphorylation in rat liver mitochondria has been studied.

The obtained data were compared with DNA intercalating properties of alkaloids detected by the method of thermodenaturation (DNA melting curve plots). It was found that chelerythrine and sanguinarine blocked absorption and accumulation of calcium cations and inhibited oxidative phosphorylation, while the coptisine significantly diminished those indices. Chelidonine, colchicines and colchamine had no influence on the studied characteristics. The effect of alkaloids upon mitochondria functional state correlated tightly with their DNA intercalating properties: chelerythrine and sanguinarine were strong intercalators, while coptisine was a weak one, and chelidonine, colchicine and colchamine did not interact with DNA and caused no changes in its melting point.

Correlation coefficient between the intercalating properties of alkaloids and their inhibition of calcium accumulation was.89, and with their oxidative phosphorylation inhibition -.93. It is suggested that the effect of studied alkaloids upon functional properties of mitochondria can be mediated by mtDNA. DNA topoisomerase I is essential for cellular metabolism and survival. It is also the target of a novel class of anticancer drugs active against previously refractory solid tumors, the camptothecins. The present review describes the topoisomerase I catalytic mechanisms with particular emphasis on the cleavage complex that represents the enzyme's catalytic intermediate and the site of action for camptothecins. Roles of topoisomerase I in DNA replication, transcription and recombination are also reviewed. Because of the importance of topoisomerase I as a chemotherapeutic target, we review the mechanisms of action of camptothecins and the other topoisomerase I inhibitors identified to date.

The DNA intercalating agents 4'-(9-acridinyl-amino) methanesulfon-m-anisidide (m-AMSA) and adriamycin were studied by using filter elution methods to measure DNA single-strand breaks (SSB's), DNA-protein cross-links (DPC's), and double-stranded breaks (DSB's) in mouse leukemia L1210 cells. Both compounds produced SSB's and DPC's at nearly 1:1 ratios. The SSB's and DPC's were shown to be localized with respect to each other; this was inferred from the finding that filter assays based on protein adsorption completely prevented the elution of the DNA single-strand segments between SSB's. In the case of m-AMSA, which produces relatively high frequencies of DNA lesions, the possibility that a protein bridges across the SSB was excluded by alkaline sedimentation studies.

The o-AMSA isomer is much less cytotoxic than m-AMSA and did not produce protein-associated strand breaks. The simplest model to explain the results is that a protein becomes covalently bound to either the 3' or the 5' termini of the intercalator-induced strand breaks.

At moderately cytotoxic doses, m-AMSA yielded much larger frequencies of protein-associated SSB's than did adriamycin. m-AMSA-induced protein-associated SSB's saturated at approximately 60000 per cell over a concentration range in which m-AMSA uptake by the cells was proportional to the drug concentration. m-AMSA was found to enter and exit from cells very rapidly at 37 degrees C; protein-associated SSB's and DSB's also appeared and disappeared rapidly. At reduced temperature, however, the appearance and disappearance of protein-associated SSB's could be blocked while m-AMSA entry and exit still occurred.

The saturation behaviour and temperature dependence suggest that the formation and disappearance of protein-associated strand breaks is enzymatic. The simplest hypothesis is that the linked protein is a nuclease, such as a topoisomerase, which becomes bound to one terminus of the strand break it produces. It is proposed that topoisomerases producing SSB's and DSB's are stimulated to different degrees by different intercalators. Topoisomerase II mediated DNA scission induced by both a nonintercalating agent [4'-demethylepipodophyllotoxin 4-(4,6-O-ethylidene-beta-D-glucopyranoside)(VP-16)] and an intercalator [4'-(9-

acridinylamino) methanesulfon-m-anisidide (m-AMSA)] wasstudied as a function of proliferation in Chinese hamster ovary (CHO), HeLa, and mouse leukemia L1210 cell lines. Log-phase CHO cells exhibited dose-dependent drug-induced DNA breaks, while plateau cells were found to be resistant to the effects of VP-16 and m-AMSA. Neither decreased viability nor altered drug uptake accounted for the drug resistance of these confluent cells. In contrast to CHO cells, plateau-phase HeLa and L1210 cells remained sensitive to VP-16 and m-AMSA. Recovery of drug sensitivity by plateau-phase CHO cells was found to reach a maximum approximately 18 h after these cells regained exponential growth and was independent of DNA synthesis.

DNA strand break frequency correlated with cytotoxicity in CHO cells; log cells demonstrated an inverse log linear relationship between drug dose (or DNA damage) and colony survival, whereas plateau-derived colony survival was virtually unaffected by increasing drug dose. Topoisomerase II activity, whether determined by decatenation of kinetoplast DNA, by cleavage of pBR322 DNA, or by precipitation of the DNA-topoisomerase II complex, was uniformly severalfold greater in log-phase CHO cells compared to plateau-phase cells.

The biochemical characteristics of the formation and disappearance of intercalator-induced DNA double-strand breaks (DSB) were studied in nuclei from mouse leukemia L1210 cells by using filter elution methodology [Bradley, M. O.,& Kohn, K.W.(1979) Nucleic Acids Res. 7, 793-804]. The three intercalators used were 4'-(9-acridinylamino)-methanesulfon-m-anisidide (m-AMSA), 5-iminodaunorubicin (5-ID), and ellipticine. These compounds differ in that they produced predominantly DNA single-strand breaks (SSB)(m-AMSA) or predominantly DNA double-strand breaks (ellipticine) or a mixture of both SSB and DSB (5-ID) in whole cells. In isolated nuclei, each intercalator produced DSB at a frequency comparable to that which is produced in whole cells. Moreover, these DNA breaks reversed within 30 min after drug removal. It thus appeared that neither ATP nor other nucleotides were necessary for intercalator-dependent DNA nicking-closing reactions. The formation of the intercalator-induced DSB was reduced at ice temperature. Break formation was also reduced in the absence of magnesium, at a pH above 6.4 and at NaCl concentrations above 200 mM. In the presence of ATP and ATP analogues, the intercalator-induced cleavage was enhanced. These results suggest that the intercalator-induced DSB are enzymatically mediated and that the enzymes involved in these reactions can catalyze DNA double-strand cleavage and rejoining in the absence of ATP, although the occupancy of an ATP binding site might convert the enzyme to a form more reactive to intercalators. Three inhibitors of DNA topoisomerase II—novobiocin, nalidixic acid, and norfloxacin—reduced the formation of DNA strand breaks.

MECHANISMS OF ALKYLATING AGENTS

Alkylating agents work by three different mechanisms all of which achieve the same end result - disruption of DNA function and cell death.In the first mechanism an alkylating agent (represented in the figure below as a pink star) attaches alkyl groups (small carbon compounds-depicted as pink triangles) to DNA bases. This alteration results in the DNA being fragmented by repair enzymes in their attempts to replace the alkylated bases (frame 3 of the diagram below). Alkylated bases prevent DNA synthesis and RN transcription from the affected DNA.

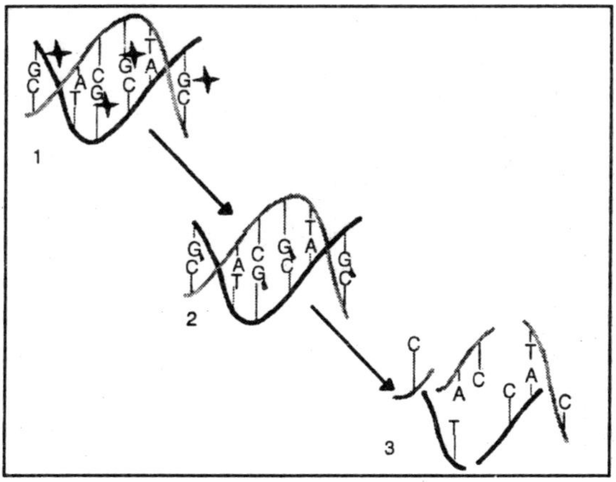

A second mechanism by which alkylating agents cause DNA damage is the formation of cross-bridges, bonds between atoms in the DNA (pink linkages below). In this process, two bases are linked together by an alkylating agent that has two DNA binding sites. Bridges can be formed within a single molecule of DNA (as shown below) or a cross-bridge may connect two different DNA molecules. Cross-linking prevents DNA from being separated for synthesis or transcription.

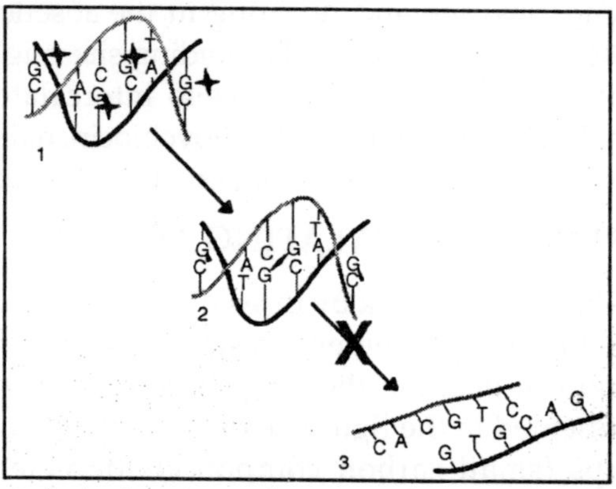

The third mechanism of action of alkylating agents is the induction of mispairing of the nucleotidesleading to mutations. In a normal DNA double helix, A always pairs with (is across from) T and G always pairs with C. As the figure below shows, alkylated G bases may erroneously pair with Ts. If this altered pairing is not corrected it may lead to a permanent mutation.

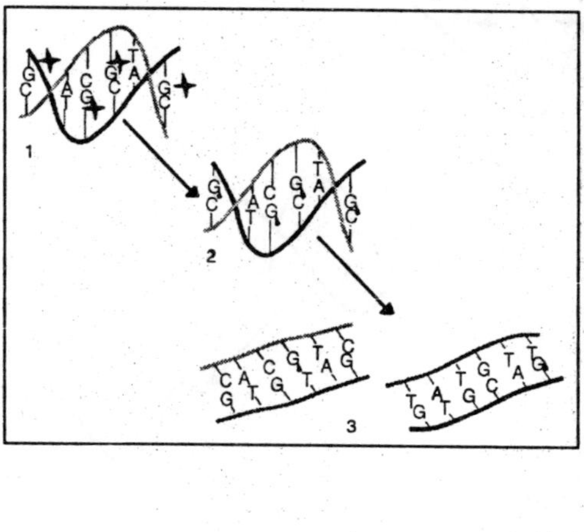

Alkylating agents are one of the earliest and most commonly used chemotherapy agents used for cancer treatments. Their use in cancer treatments started in early 1940s. Majority of alkaline agents are active or dormant nitrogen mustards, which are poisonous compound initially used for certain military purposes. Chlorambucil, Cyclophosphamide, CCNU, Melphalan, Procarbazine, Thiotepa, BCNU, and Busulfan are some of the commonly used alkylating agents.

Although they might differ in their clinical activity, action mechanism of all alkylating agents is the same. These agents work directly on the DNA and prevent the cell division process by cross-linking and breaking the DNA strands and causing abnormal base pairing. When a DNA is altered in this manner, undesired cellular activity comes to a halt and the cell dies eventually.

Alkylating chemotherapy drugs are effective during all phases of cell cycle. Therefore, they are used to treat a large number of cancers. However, they are more effective in treating slow-growing cancers such as solid tumors and leukemia. Long term use of alkylating agents can lead to permanent infertility by decreasing sperm production in males, and causing menstruation cessation in females. Many alkylating agents can also lead to secondary cancers such as Acute Myeloid Leukemia, years after the therapy.

ANTIMETABOLITES

Structure of antimetabolites (antineoplastic agents) is similar to certain compounds such as vitamins, amino acids, and precursors of DNA or RNA, found naturally in human body. Antimetabolites help in treatment cancer by inhibiting cell division thereby hindering the growth of tumor cells. These agents get incorporated in the DNA or RNA to interfere with the process of division of cancer cells.

Antimetabolites were first discovered and it is found that folic acid analog can reduce childhood leukemia. Out of 16 patients he tested, 10 displayed hematologic improvement. This discovery laid the foundation that enabled scientist to synthesize many new agents that could inhibit biological enzymatic reactions. Antimetabolites are found to be useful in treating chronic and acute cases of leukemia and various tumors. They are commonly used to treat gastrointestinal tract, breast, and ovary tumors.

Methotraxate, which is a commonly used antimetabolites chemotherapy agent, is effective in the S-phase of the cell cycle. It works by inhibiting an enzyme that is essential for DNA synthesis. 6-mercaptopurine and 5-fluorouracil (5FU) are two other commonly used antimetabolites. 5-Fluorouracil (5-FU) works by interfering with the DNA components, nucleotide, to stop DNA synthesis. This drug is used to treat many different types of cancers including breast, esophageal, head, neck, and gastric cancers. 6-mercaptopurine is an analogue of hypoxanthine and is commonly used to treat Acute Lymphoblastic Leukemia (ALL). Other popular antimetabolite chemotherapy drugs are Thioguanine, Cytarabine, Cladribine. Gemcitabine, and Fludarabine.

ANTHRACYCLINES

Anthracyclines were developed between 1970s and 1990s and are daunosamine and tetra-hydronaphthacenedione-based chemotherapy agents. These compounds are cell-cycle nonspecific and are used to treat a large number of cancers including lymphomas, leukemia, and uterine, ovarian, lung and breast cancers.

Anthracyclines drugs are developed from natural resources. For instance, daunorubicin is developed by isolating it from soil-dwelling fungus Streptomyces. Similarly, Doxorubicin, which is another commonly used anthracycline chemotherapy agent, is isolated from mutated strain of Streptomyces. Although both the drugs have similar clinical action mechanisms, doxorubicin is more effective in treating solid tumors. Idarubicin, Epirubicin, and Mitoxantrone are few of the other commonly used anthracycline chemotherapy drugs. Anthracyclines work by forming free oxygen radicals that breaks DNA strands thereby inhibiting DNA synthesis and function. These chemotherapeutic agents form a complex with DNA and enzyme to inhibit the topoisomerase enzyme. Topoisomerase is an enzyme class that causes the supercoiling of DNA, allowing DNA repair, transcription, and replication.

One of the main side effects of anthracyclines is that it can damage cells of heart muscle along with the DNA of cancer cell leading to cardiac toxicity.

ANTITUMOR ANTIBIOTICS

Antitumor antibiotics are also developed from the soil fungus Streptomyces. These drugs are widely used to treat and suppress development of tumors in the body. Similar to anthracyclines, antitumor antibiotics drugs also form free oxygen radicals that result in DNA strand breaks, killing the growth of cancer cells. In most of the cases, these drugs are used in combination with other chemotherapy agents. Bleomycin is one of the commonly used antitumor antibiotic used to treat testicular cancer and hodgkin's lymphoma. The most serious side effect of this drug is lung toxicity that occurs when the oxygen radical formed by the antitumor antibiotics damages lung cells along with the cancer cells.

MONOCLONAL ANTIBODIE

Monoclonal antibodies are one of the newer chemotherapy agents approved for cancer treatment by the Food and Drug Administration (FDA) in 1997. Alemtuzumab (Campath), Bevacizumab (Avastin), Cetuximab (Erbitux), Gemtuzumab (Mylotarg), Ibritumomab (Zevalin), Panitumumab (Vectibix), Rituximab (Rituxan), Tositumomab (Bexxar), and Trastuzumab (Herceptin) are some of the FDA approved monoclonal drugs used in chemotherapeutic cancer treatments. The treatment is known to be useful in treating colon, lung, head, neck, and breast cancers. Some of the monoclonal drugs are used to treat chronic lymphocytic leukemia, acute myelogenous leukemia, and non-Hodgkin's lymphoma. Monoclonal antibodies work by attaching to certain parts of the tumor-specific antigens and make them easily recognizable by the host's immune system. They also prevent growth of cancer cells by blocking the cell receptors to which chemicals called 'growth factors' attach promoting cell growth. Monoclonal antibodies can be combined with radioactive particles and other powerful anticancer drugs to deliver them directly to cancer cells. Using this method, long term radioactive treatment and anticancer drugs can be given to patients without causing any serious harm to other healthy cells of the body.

PLATINUMS

Platinum-based natural metal derivatives were found to be useful for cancer treatments around 150 years ago with the synthesis of cisplatin. However, there clinical use did not commence until 30 years ago. Platinum-based chemotherapy agents work by cross-linking subunits of DNA. These agents act during any part of cell cycle and help in treating cancer by impairing DNA synthesis, transcription, and function.

Cisplatin, although found to be useful in treating testicular and lung cancer, is highly toxic and can severely damage the kidneys of the patient. Second generation platinum-complex carboplatin is found to be much less toxic in comparison to cisplatin and has fewer kidney-related side effects. Oxaliplatin, which is third generation platinum-based complex, is found to be helpful in treating colon cancer. Although, oxaliplatin does not cause any toxicity in kidney it can lead to severe neuropathies.

PLANT ALKALOIDS

Plant alkaloid chemotherapy agents, as the name suggests, are plant derivatives. They are cell-specific chemotherapy agents. However, the cycle affected is based on the drug used for the treatment. They are primarily categorized into four groups: topoisomerase inhibitors, vinca alkaloids, taxanes, and epipodophyllotoxins. Plant alkaloids are cell-cycle specific, but the cycle affected varies from drug to drug. Vincristine (Oncovin) is a plant alkaloid of interest in mesothelioma treatment.

TOPOISOMERASE INHIBITORS

Topoisomerase inhibitors are chemotherapy agents are

categorized into Type I and Type II Topoisomerases inhibitors and they work by interfering with DNA transcription, replication, and function to prevent DNA supercoiling.

- *Type I Topoisomerase inhibitors:* These chemotherapy agents are extracted from the bark and wood of the Chinese tree Camptotheca accuminata. They work by forming a complex with topoisomerase DNA. This in turn suppresses the function of topoisomerase. Camptothecins which includes irinotecan and topotecan are commonly used type I topoisomerase inhibitors, first discovered in the late 1950s.
- *Type II Topoisomerase inhibitors:* These are extracted from the alkaloids found in the roots of May Apple plants. They work in the in the work in the late S and G2 phases of the cell cycle.
  Amsacrine, etoposide, etoposide phosphate, and teniposide are some of the examples of type II topoisomerase inhibitors.

VINCA ALKALOIDS

Vinca alkaloids are derived from the periwinkle plant, Vinca rosea (Catharanthus roseus) and are known to be used by the natives of Madagascar to treat diabetes.

Although not useful in controlling diabetes, vinca alkaloids, are useful in treating leukemias. They are effective in the M phase of the cell cycle and work by inhibiting tubulin assembly in microtubules. Vincristine, Vinblastine, Vinorelbine, and Vindesine are some of the popularly used vinca alkaloid chemotherapy agents used today. Major side effect of vinca alkaloids is that they can cause neurotoxicity in patients.

TAXANES

Taxanes are plant alkaloids that were first developed in 1963 by isolating it from first isolated from the bark of the Pacific yew tree, Taxus brevifolia in 1963. Paclitaxel, which is the active components of taxanes was first discovered in 1971 and was made available for clinical use in the year 1993. Taxanes also work in the M-phase of the cell cycle and inhibit the function of microtubules by binding with them. Paclitaxel and docetaxel are commonly used taxanes. Taxanes chemotherapy agents are used to treat a large array of cancers including breast, ovarian, lung, head and neck, gastric, esophageal, prostrate and gastric cancers. The main side effect of taxanes is that they lower the blood counts in patients.

ANTI-CANCER DRUG

INTRODUCTION

The available anticancer drugs have distinct mechanisms of action which may vary in their effects on different types of normal and cancer cells. A single "cure" for cancer has proved elusive since there is not a single type of cancer but as many as 100 different types of cancer. In addition, there are very few demonstrable biochemical differences between cancerous cells and normal cells. For this reason the effectiveness of many anticancer drugs is limited by their toxicity to normal rapidly growing cells in the intestinal and bone marrow areas. A final problem is that cancerous cells which are initially suppressed by a specific drug may develop a resistance to that drug. For this reason cancer chemotherapy may consist of using several drugs in combination for varying lengths of time.

CANCER CHEMOTHERAPY

Chemotherapy drugs, are sometimes feared because of a patient's concern about toxic effects. Their role is to slow and hopefully halt the growth and spread of a cancer. There are three goals associated with the use of the most commonly-used anticancer agents.

- Damage the DNA of the affected cancer cells.
- Inhibit the synthesis of new DNA strands to stop the cell from replicating, because the replication of the cell is what allows the tumor to grow.
- Stop mitosis or the actual splitting of the original cell into two new cells. Stopping mitosis stops cell division (replication) of the cancer and may ultimately halt the progression of the cancer.

Unfortunately, the majority of drugs currently on the market are not specific, which leads to the many common side effects associated with cancer chemotherapy. Because the common approach of all chemotherapy is to decrease the growth rate (cell division) of the cancer cells, the side effects are seen in bodily systems that naturally have a rapid turnover of cells iincluding skin, hair, gastrointestinal, and bone marrow. These healthy, normal cells, also end up damaged by the chemotherapy programme.

Catagories of Chemotherapy Drugs

In general, chemotherapy agents can be divided into three main categories based on their mechanism of action.

Stop the synthesis of pre DNA molecule building blocks: These agents work in a number of different ways. DNA building blocks are folic acid, heterocyclic bases, and nucleotides, which are made naturally within cells. All of these agents work to block some step in the formation of nucleotides or deoxyribonucleotides (necessary for making DNA). When these steps are blocked, the nucleotides, which arethe building blocks of DNA and RNA, can not be synthesized. Thus the cells can not replicate because they can nnot make DNA without the nucleotides.

Examples of drugs in this class include:

- Methotrexate (Abitrexate®),
- Fluorouracil (Adrucil®),
- Hydroxyurea (Hydrea®), and
- Mercaptopurine (Purinethol®).

Directly damage the DNA in the nucleus of the cell:

These agents chemically damage DNA and RNA. They disrupt replication of the DNA and either totally halt replication or cause the manufacture of nonsense DNA or RNA (*i.e.* the new DNA or RNA does not code for anything useful).

Examples of drugs in this class include cisplatin (Platinol®) and antibiotics - daunorubicin (Cerubidine®), doxorubicin (Adriamycin®), and etoposide (VePesid®).

Effect the synthesis or breakdown of the mitotic spindles:

Mitotic spindles serve as molecular railroads with "North and South Poles" in the cell when a cell starts to divide itself into two new cells. These spindles are very important because they help to split the newly copied DNA such that a copy goes to each of the two new cells during cell division. These drugs disrupt the formation of these spindles and therefore interrupt cell division.

Examples of drugs in this class of miotic disrupters include: Vinblastine (Velban), Vincristine (Oncovin) and Pacitaxel (Taxol).

Methotrexate

Methotrexate inhibits folic acid reductase which is responsible for the conversion of folic acid to tetrahydrofolic acid. At two stages in the biosynthesis of purines (adenine and guanine) and at one stage in the synthesis of pyrimidines (thymine, cytosine, and uracil), one-carbon transfer reactions occur which require specific coenzymes synthesized in the cell from tetrahydrofolic acid.

Tetrahydrofolic acid itself is synthesized in the cell from folic acid with the help of an enzyme, folic acid reductase. Methotrexate looks a lot like folic acid to the enzyme, so it binds to it thinking that it is folic acid. In fact, methotrexate looks so good to the enzyme that it binds to it quite strongly and inhibits the enzyme. Thus, DNA synthesis cannot proceed because the coenzymes needed for one-carbon transfer reactions are not produced from tetrahydrofolic acid because there is no tetrahydrofolic acid. Again, without DNA, no cell division.

5-Fluorouracil

5-Fluorouracil (5-FU; Adrucil, Fluorouracil, Efudex, Fluoroplex) is an effective pyrimidine antimetabolite. Fluorouracil is synthesized into the nucleotide, 5-fluoro-2-deoxyuridine. This product acts as an antimetabolite by inhibiting the synthesis of 2-deoxythymidine because the carbon - fluorine bond is extremely stable and prevents the addition of a methyl group in the 5-position. The failure to synthesize the thymidine nucleotide results in little or no production of DNA.

Two other similar drugs include: gemcitabine (Gemzar) and arabinosylcytosine (araC). They all work through similar mechanisms.

Hydroxyurea

Hydroxyurea blocks an enzyme which converts the cytosine nucleotide into the deoxy derivative. In addition, DNA synthesis is further inhibited because hydroxyurea blocks the incorporation of the thymidine nucleotide into the DNA strand.

Mercaptopurine

Mercaptopurine, a chemical analog of the purine adenine, inhibits the biosynthesis of adenine nucleotides by acting as an antimetabolite. In the body, 6-MP is converted to the corresponding ribonucleotide. 6-MP ribonucleotide is a potent inhibitor of the conversion of a compound called inosinic acid to adenine Without adenine, DNA cannot be synthesized. 6-MP also works by being incorporated into nucleic acids as thioguanosine, rendering the resulting nucleic acids (DNA, RNA) unable to direct proper protein synthesis. Adenine and Mercaptopurine - Chime in new window

Thioguanine

Thioguanine is an antimetabolite in the synthesis of guanine nucleotides.

Alkylating Agents

Alkylating agents involve reactions with guanine in DNA. These drugs add methyl or other alkyl groups onto molecules where they do not belong. This in turn inhibits their correct utilization by base pairing and causes a miscoding of DNA.

In the first mechanism an alkylating agent attaches alkyl groups to DNA bases. This alteration results in the DNA being fragmented by repair enzymes in their attempts to replace the alkylated bases. A second mechanism by which alkylating agents cause DNA damage is the formation of cross-bridges, bonds between atoms in the DNA. In this process, two bases are linked together by an alkylating agent that has two DNA binding sites. Cross-linking prevents DNA from being separated for synthesis or transcription. The third mechanism of action of alkylating agents causes the mispairing of the nucleotides leading to mutations.

There are six groups of alkylating agents: nitrogen mustards; ethylenimes; alkylsulfonates; triazenes; piperazines; and nitrosureas. Cyclosporamide is a classical example of the role of the host metabolism in the activation of an alkylating agent and is one or the most widely used agents of this class. It was hoped that the cancer cells might posses enzymes capable of accomplishing the cleavage, thus resulting in the selective production of an activated nitrogen mustard in the malignant cells. Compare the top and bottom structures in the graphic on the left.

Antibiotics

A number of antibiotics such as anthracyclines, dactinomycin, bleomycin, adriamycin, mithramycin, bind to DNA and inactivate it. Thus the synthesis of RNA is prevented.

General properties of these drugs include: interaction with DNA in a variety of different ways including intercalation (squeezing between the base pairs), DNA strand breakage and inhibition with the enzyme topoisomerase II. Most of these compounds have been isolated from natural sources and antibiotics. However, they lack the specificity of the antimicrobial antibiotics and thus produce significant toxicity.

The anthracyclines are among the most important antitumor drugs available. Doxorubicin is widely used for the treatment of several solid tumors while daunorubicin and idarubicin are used exclusively for the treatment of leukemia.

These agents have a number of important effects including: intercalating (squeezing between the base pairs) with DNA affecting many functions of the DNA including DNA and RNA synthesis. Breakage of the DNA strand can also occur by inhibition of the enzyme topoisomerase II.

Dactinomycin (Actinomycin D): At low concentrations dactinomycin inhibits DNA directed RNA synthesis and at higher concentrations DNA synthesis is also inhibited. All types of RNA are affected, but ribosomal RNA is more sensitive. Dactinomycin binds to double stranded DNA, permitting RNA chain initiation but blocking chain elongation. Binding to the DNA depends on the presence of guanine.

Mitotic Disrupters

Plant alkaloids like vincristine prevent cell division, or mitosis. There are several phases of mitosis, one of which is the metaphase. During metaphase, the cell pulls duplicated DNA chromosomes to either side of the parent cell in structures called "spindles". These spindles ensure that each new cell gets a full set of DNA. Spindles are microtubular fibres formed with the help of the protein "tubulin". Vincristine binds to tubulin, thus preventing the formation of spindles and cell division. In contrast to other microtubule antagonists, taxol disrupts the equilibrium between free tubulin and mircrotubules by shifting it in the direction of assembly, rather than disassembly. As a result, taxol treatment causes both the stabilization of microtubules and the formation of abnormal bundles of microtubules. The net effect is still the disruption of mitosis.

RIBONUCLEIC ACID

Ribonucleic acid (RNA) is a biologically important type of molecule that consists of a long chain of nucleotide units. Each nucleotide consists of a nitrogenous base, a ribose sugar, and a phosphate. RNA is very similar to DNA, but differs in a few important structural details: in the cell, RNA is usually single-stranded, while DNA is usually double-stranded; RNA nucleotides contain ribose while DNA contains deoxyribose (a type of ribose that lacks one oxygen atom); and RNA has the base uracil rather than thymine that is present in DNA.

Ribonucleic acid (RNA) is a linear molecule composed of four types of smaller molecules called ribonucleotide bases: adenine (A), cytosine (C), guanine (G), and uracil (U). RNA is often compared to a copy from a reference book, or a template, because it carries the same information as its DNA template but is not used for long-term storage. Each ribonucleotide base consists of a ribose sugar, a phosphate group, and a nitrogenous base. Adjacent ribose nucleotide bases are chemically attached to one another in a chain via chemical bonds called phosphodiester bonds. Unlike DNA, RNA is usually single-stranded. Additionally, RNA contains ribose sugars rather than deoxyribose sugars, which makes RNA more unstable and more prone to degradation. RNA is synthesized from DNA by an enzyme known as RNA polymerase during a process called transcription. The new RNA sequences are complementary to their DNA template, rather than being identical copies of the template. RNA is then translated into proteins by structures called ribosomes. There are three types of RNA involved in the translation process: messenger RNA (mRNA), transfer RNA (tRNA), and ribosomal RNA (rRNA). Although some RNA molecules are passive copies of DNA, many play crucial, active roles in the cell. For example, some RNA molecules are involved in switching genes on and off, and other RNA molecules make up the critical protein synthesis machinery in ribosomes.

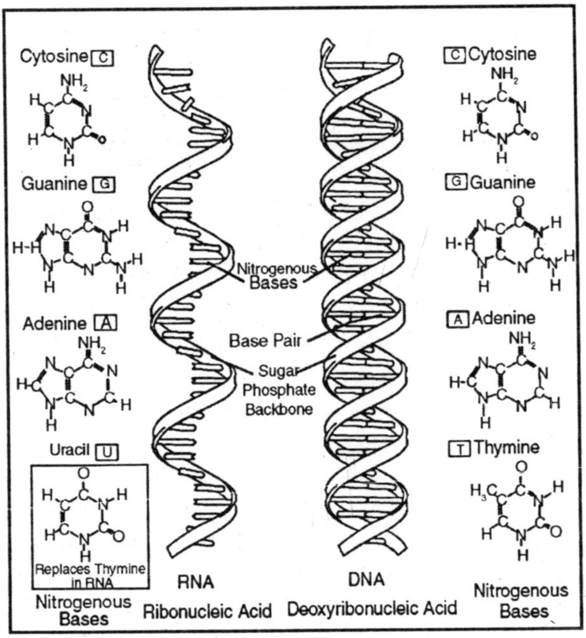

RNA is transcribed from DNA by enzymes called RNA polymerases and is generally further processed by other enzymes. RNA is central to the synthesis of proteins. Here, a type of RNA called messenger RNA carries information from DNA to structures called ribosomes.

These ribosomes are made from proteins and ribosomal RNAs, which come together to form a molecular machine that can read messenger RNAs and translate the information they carry into proteins. There are many RNAs with other roles – in particular regulating which genes are expressed, but also as the genomes of most viruses.

MESSENGER RNA (MRNA) AND ITS PRECURSORS

The primary RNA transcripts are produced by copying the DNA of a protein-encoding gene.[1] Subsequent processing steps include addition, removal, and modification of nucleotides as well as splicing events that excise internal segments. The mature mRNA is translated to produce a protein whose amino acid sequence is determined by the sequence of the coding region in the gene. The messenger RNA, as the name implies, is the molecule that carries the message from the gene to the protein synthesis machinery.

RIBOSOMAL RNA (RRNA)

The ribosomes are the most important part of the translation machinery and it has long been known that much of the mass of ribosomes is due to the presence several types of ribosomal RNA. These are noncoding RNAs produced by transcription of ribosomal RNA genes.[3] One of the key steps in translation—formation of the peptide bond—is catalyzed by the rRNA component of the ribosome. It is the major catalytic RNA in cells.

TRANSFER RNA (TRNA)

tRNAs are intermediates in protein synthesis. There are many different tRNA molecules in every cell and each one binds a specific amino acid, yielding an aminoacylated-tRNA (aa-tRNA). Each different aminoacylated tRNA interacts with a particular codon in mRNA thus delivering the correct amino acid to to the site of protein synthesis.

SMALL RNAS

The small RNAs represent a heterogeneous category of RNAs covering a wide ranges of functions. Some of them have catalytic functions—RNAse P is the classic example. Some of them are structural components of ribonucleoprotein complexes (*e.g.* signal recognition particle). Some of them are guide RNAs involved in various processing events. The best known examples of guide RNAs are the small RNAs of the spliceosome complexes that mediate the splicing of mRNA precursors.[2]Other small RNAs were known to be involved in the regulation of gene expression.

GENOMIC RNA

Some viruses, notably retroviruses, have an RNA genome instead of a DNA genome. In addition, the mobility of various transposons is due to an intermediate RNA copy of the transposon sequence (retrotransposons).

This was the state of knowledge 25 years ago. Since then, the study of RNA has made remarkable progress. Our knowledge of all the fundamental processes—transcription, processing, and catalysis—has expanded enormously.

The biggest change is in the area of small RNAs. Today there are several categories of small RNAs—siRNA, microRNA, piRNA—that were only discovered in the past 10-15 years. The functions of these small RNA molecules are still being worked out. There's little doubt that some of them have important biological roles but there's considerable controversy over what percentage might be artifacts of one sort or another.

RNA replaces DNA as the genetic material. As with DNA, RNA follows specific base pairing rules, except that in RNA the base uracil replaces the base thymine (*i.e.,* instead of an adenine-thymine or A-T pairing, there is an adenine-uracil or A-U pairing). Accordingly, when RNA acts as a carrier of genetic information, uracil replaces thymine in the genetic code.

In humans, messenger RNA (mRNA) is the product of transcription and acts to convey genetic information from the nucleus to the protein assembly complex at the ribosome. The ribsome is composed of ribosomal RNA (rRNA) and other proteins. Transfer RNAs (tRNA) act to catalyze the translation process by acting as carriers of specific amino acids. Because tRNAs bind to specific sites on the strand of mRNA, the sequence of amino acids subsequently inserted into the synthesized protein is both specific and genetically determined by the nucleotide sequence in DNA from which the mRNA strand was originally transcribed.

Other forms of RNA perform important roles in other biochemical reactions. Regardless of function, RNA is a biopolymer made up of ribonucleotide units and is present in all living cells and some viruses. The chemical units of RNA are ribonucleotide monomers consisting of a ribose sugar ($C_5H_{10}O_5$) phosphorylated at the third carbon (C3) and linked to one of four bases through a type of chemical linkage formed between a sugar and a base by a condensation reaction (glycosidic bond). The four bases found in RNA are adenine (A), guanine (G), cytosine (C), and uracil (U). Other bases may also be found, although they are generally modified versions of these four (*e.g.*, methylated bases are found in parts of tRNA).

The single nucleotides (monomers) of RNA form a linear chain by linking their phosphate groups and sugars in phosphodiester bonds. RNA does not form a double stranded alpha-helix as does DNA. In some parts of the RNA molecule, there is folding into alpha-helical-like regions. Corresponding to their unique functions, messenger RNA (mRNA), ribosomal RNA (rRNA), and transfer RNA (tRNA) all have different three-dimensional structures. In higher eukaryotic organisms, different RNAs are found distributed throughout the cell—in the nucleus, cytoplasm, and also in cytoplasmic organelles such as mitochondria and, in plants, chloroplasts.

The nucleus is the chief site of RNA synthesis and the source of all cytoplasmic RNA, while mitochondria and chloroplasts synthesize their RNA from their own DNA. rRNA is synthesized by the nucleoli within the nucleus, while the high molecular weight precursor to cytoplasmic mRNA, sometimes termed heterogeneous nuclear or hnRNA, is transcribed on the DNA chromatin. Low molecular weight RNA also occurs in the nucleus and consists partly of tRNA and partly of RNA, which has a regulatory function in geneactivation. The cytoplasm contains tRNA and rRNA in the ribosomes and mRNA in polysomes, or polyribosomes. The latter are the structural units of protein biosynthesis, consisting of several ribosomes attached to a strand of mRNA.

The function of mRNA is to transcribe the information held in DNA. In the cells of eukaryotic organisms, the first transcriptional product is the long, heterogenous nuclear RNA, or hnRNA. This contains both the nucleotide sequences eventually transcribed into polypeptides and large tracts of sequences not translated. Non-translated sequences are termed introns (or intervening sequences). Removal of introns, and other untranslated portions of the molecule, edits hnRNA into mRNA molecules. After editing removes as much as 90% of hnRNA, the resulting mRNA molecules are transported into the cytoplasm.

rRNA is located within ribosomes, the sites of protein biosynthesis. Ribosomes are large ellipsoid cytoplasmic organelles consisting of RNA and protein.

tRNA, the smallest known functional RNA, is essential for protein biosynthesis. Its purpose is to transfer a specific amino acid from the cytoplasm and incorporate it into the growing polypeptide chain on the polysome. Different tRNAs contain between 70 and 85 nucleotides. The most characteristic feature of tRNA is that it contains the anticodon, a sequence of three nucleotides specific for the mRNA codon sequence. There is at least one tRNA per cell bearing the anticodon for each of the 20 amino acids. The aminoacyl-tRNA (the tRNA carrying the amino acid) binds to the large subunit of a ribosome, where antiparallel basepairing occurs between the anticodon of the tRNA and the complementary codon of the associated mRNA. The specificity of this base pairing ensures that the amino acid inserts into the correct position in the growing protein polypeptide chain. During translation, the deacylated tRNA (*i.e.*, with its amino acid removed) is released from the ribosome and becomes available once again for recharging with its amino acid.

DNA-dependent RNA synthesis is the process of RNA sythesis on a template of DNA. According to the rules of base pairing, the base sequence of DNA determines the synthesis of a complementary base sequence in RNA. Assisted (catalyzed) by the enzyme RNA polymerase, the growing RNA chain releases from the template so that the process can start again, even before the previous molecule is complete. Termination codons and a termination factor known as rho-factor end the synthesis process. In certain viruses, RNA-dependent RNA synthesis occurs, with the viral RNA acting as a template for the synthesis of new RNA.

Each nucleotide in RNA contains a ribose sugar, with carbons numbered 1' through 5'. A base is attached to the 1' position, generally adenine (A), cytosine (C), guanine (G) or uracil (U). Adenine and guanine are purines, cytosine and uracil are pyrimidines. A phosphate group is attached to the 3' position of one ribose and the 5' position of the next. The phosphate groups have a negative charge each at physiological pH, making RNA a charged molecule (polyanion). The bases may form hydrogen bonds between cytosine and guanine, between adenine and uracil and between guanine and uracil.However other interactions are possible, such as a group of adenine bases binding to each other in a bulge, or the GNRA tetraloop that has a guanine–adenine base-pair.

An important structural feature of RNA that distinguishes it from DNA is the presence of a hydroxyl group at the 2' position of the ribose sugar. The presence of this functional group causes the helix to adopt the A-form geometry rather than the B-form most commonly observed in DNA. This results in a very deep and narrow major groove and a shallow and wide minor groove. A second consequence of the presence of the 2'-hydroxyl group is that in conformationally flexible regions of an RNA molecule (that is, not involved in formation of a double helix), it can chemically attack the adjacent phosphodiester bond to cleave the backbone. RNA is transcribed with only four bases (adenine, cytosine, guanine and uracil), but there are numerous modified bases and sugars in mature RNAs. Pseudouridine (Ø), in which the linkage between uracil and ribose is changed from a C–N bond to a C–C bond, and ribothymidine, are found in various places (most notably in the TØC loop of tRNA).Another notable modified base is hypoxanthine, a deaminated adenine base whose nucleoside is called inosine.

Inosine plays a key role in the wobble hypothesis of the genetic code.There are nearly 100 other naturally occurring modified nucleosides,of which pseudouridine and nucleosides with 2'-O-methylribose are the most common.The specific roles of many of these modifications in RNA are not fully understood. However, it is notable that in ribosomal RNA, many of the post-transcriptional modifications occur in highly functional regions, such as the peptidyl transferase center and the subunit interface, implying that they are important for normal function.

RNA and DNA are both nucleic acids, but differ in three main ways. First, unlike DNA which is double-stranded, RNA is a single-stranded molecule in most of its biological roles and has a much shorter chain of nucleotides. Second, while DNA contains deoxyribose, RNA contains ribose, (there is no hydroxyl group attached to the pentose ring in the 2' position in DNA). These hydroxyl groups make RNA less stable than DNA because it is more prone to hydrolysis. Third, the complementary base to adenine is not thymine, as it is in DNA, but rather uracil, which is an unmethylated form of thymine.Like DNA, most biologically active RNAs, including mRNA, tRNA, rRNA, snRNAs and other non-coding RNAs, contain self-complementary sequences that allow parts of the RNA to fold and pair with itself to form double helices. Structural analysis of these RNAs have revealed that they are highly structured.

Bibliography

Adelberg: *Papers on Bacterial Genetics,* Boston: Little, Brown and Company, 2001.

Ausubel, F.: *Current Protocols in Molecular Biology,* New York: John Wiley and Sons Press, 2000.

Broach, J.R.: *The Molecular Biology of the Yeast,* New York: Worth Publishers, 2005.

Brock, T.: *"Gene Manipulation and Genetic Engineering,* New York: Freeman and Company, 2006.

Bud, R.: *"Janus-Faced Biotechnology - An Historical Perspective"*, New York: Oxford Press, 2003.

Dixon: *After Man - A Zoology of the Future,* New York: Martin's Press, 2000.

Dodds, H.: *Plant Genetic Engineering,* New York: Cambridge University Press, 2004.

Goodman, D.C.: *From Farming to Biotechnology: A Theory of Agro-Industrial Development,* Oxford: Blackwell Publication, 2006.

Jensen, J.: *Karyotyping with Hypercard,* USA: The American Biology Teacher, 2000.

Micklos, D.A.: *DNA Science: A First Course in Recombinant DNA Technology,* London: Cold Spring Harbor Laboratory, 2005.

Morholt, E.: *Sourcebook for the Biological Sciences,* San Diego: Harcourt Brace JOvanovich Press, 2002.

Newcombe, H.B.: *Origin of Bacterial Variants,* London: National Association of Biology Teachers, 2003.

Retzer, W.J.: *Biotechnology Workbook,* USA: Prentice Hall Press, 2006.

Robin J.R.: *The Human Genome Project in Association with MCET the Mass LearnPike,* Cambridge: Massachusetts Publication, 2003.

Seabrook, J.: *Tremors in the Hothouse,* London: Cold Spring Harbor Laboratory, 2004.

Slater, R.J.: *Experiments in Molecular Biology,* Clifton: Humana Press, 2000.

Torrey, J.G.: *The Development of Plant Biotechnology,* New York: Oxford Press, 2001.

Towle: *Modern Biology,* Austin: Rinehart and Winston Publication, 2003.

Walden, A: *Genetic Transformation in Plants,* England: Open University Press, 2004.

Wartski, B.: *Life Science and Lego Blocks,* London: Bski International Press, 2002.